RAND NATIONAL DEFENSE RESEARCH INSTITUTE

T0306315

Assessing Aegis Program Transition to an Open-Architecture Model

Paul DeLuca, Joel B. Predd, Michael Nixon, Irv Blickstein,
Robert W. Button, James G. Kallimani, Shane Tierney

Prepared for the United States Navy

The research described in this report was prepared for the U.S. Navy. The research was conducted within the RAND National Defense Research Institute, a federally funded research and development center sponsored by the Office of the Secretary of Defense, the Joint Staff, the Unified Combatant Commands, the Navy, the Marine Corps, the defense agencies, and the defense Intelligence Community under Contract W91WAW-12-C-0030.

Library of Congress Cataloging-in-Publication Data

DeLuca, Paul.
Assessing Aegis program transition to an open-architecture model / Paul DeLuca, Joel B. Predd, Michael Nixon, Irv Blickstein, Robert W. Button, James G. Kallimani, Shane Tierney.
pages cm
Includes bibliographical references.
ISBN 978-0-8330-7879-7 (pbk. : alk. paper)
1. AEGIS (Weapons system) I. Title.

VF347.D45 2013
359.8'2519—dc23 2013008312

Cover photo from U.S. Navy, PEO Integrated Warfare Systems.

Published 2013 by the RAND Corporation
1776 Main Street, P.O. Box 2138, Santa Monica, CA 90407-2138
1200 South Hayes Street, Arlington, VA 22202-5050
4570 Fifth Avenue, Suite 600, Pittsburgh, PA 15213-2665
RAND URL: http://www.rand.org/
To order RAND documents or to obtain additional information, contact
Distribution Services: Telephone: (310) 451-7002;
Fax: (310) 451-6915; Email: order@rand.org

Preface

The U.S. Navy's Aegis program is a highly integrated combat system with anti-air warfare, ballistic missile defense, surface, subsurface, and strike roles. In order to reduce costs and enable the use of rapidly evolving commercial computing technology, the Navy is transitioning Aegis to use open-architecture (OA) software and commercial off-the-shelf (COTS) hardware.

In 2010, the Program Executive Office for Integrated Warfare Systems asked the RAND Corporation to evaluate the impact of this transition on the development, integration, and testing of upgrades to the Aegis weapon system. Of particular concern is the impact of modernization and fielding rates on the technical infrastructure of the Aegis fleet. A previous report by the same authors documented the methods and findings of that research effort, but incorporated proprietary information. This report removes all proprietary information and incorporates the most recent Navy Aegis plans.

This research was sponsored by the U.S. Navy and conducted within the Acquisition and Technology Policy Center of the RAND National Defense Research Institute, a federally funded research and development center sponsored by the Office of the Secretary of Defense, the Joint Staff, the Unified Combatant Commands, the U.S. Navy, the Marine Corps, the defense agencies, and the defense Intelligence Community.

For more information on the RAND Acquisition and Technology Policy Center, see http://www.rand.org/nsrd/ndri/centers/atp.html

or contact the director (contact information is provided on the web page).

Contents

CHAPTER THREE

CHAPTER FOUR

CHAPTER FIVE

Figures

Tables

Summary

Background

Aegis is a highly integrated combat system with anti-air warfare, ballistic missile defense, surface, subsurface, and strike roles that the U.S. Navy has installed on 84 of its ships. While the Navy wants to maintain the Aegis system as the preeminent combat system for surface combatants, this is an expensive and time-consuming endeavor. To reduce costs and enable the use of rapidly evolving commercial computing technology, the Navy is transitioning Aegis to use open-architecture (OA) software, a common source library (CSL), and commercial off-the-shelf (COTS) processors, taking advantage of their 18- to 24-month replacement cycle.

The Navy's transition from its legacy business model to the new integrated warfare systems (IWS) business model[1] may introduce new challenges and risks for the fleet and enterprise that develop and field the Aegis weapon system (AWS).[2] Under the legacy business model, the AWS used proprietary software operating on military-specification computing hardware. Upgrades to the AWS were developed every five to six years and fielded only to new-construction ships and those receiving a midlife upgrade. The IWS business model will

---

[1] The IWS business model is articulated in the Program Executive Office (PEO) Integrated Warfare Systems Acquisition Management Plan (2013).

[2] AWS refers specifically to the computer software and hardware, radar system (SPY-1), and vertical launch system onboard an Aegis ship. The additional sensors, communication systems, weapons, and countermeasures are part of the broader Aegis combat system (ACS).

use OA software operating on COTS computing hardware and will involve periodic upgrades for all ships, both new and in-service. The plan is to upgrade software through advanced capability builds (ACBs) every four years, independently of computing hardware upgrades, called technology insertions (TIs), which will occur every four years, with individual ships receiving every other upgrade.

The introduction of new capabilities into the Aegis fleet is likely to quicken over the next decade due to ballistic and cruise missile defense requirements. The Aegis fleet is the backbone of the U.S. Navy's surface fleet and will remain so for decades. Thus, it would be particularly detrimental to install improperly designed or tested combat systems on this fleet.

Purpose

This report focuses on issues related to the development, integration, and testing of upgrades to the AWS. Specifically, it attempts to answer the following three questions:

- How does the Navy currently develop, test, and field upgrades to the AWS, and how will that process change under the IWS business model?
- How does the IWS business model affect AWS modernization and fielding rates in terms of both the technical infrastructure and fleet capabilities?
- What modernization rate under the IWS business model should be recommended to the Navy to balance fleet capability, risk, and cost?

The IWS business model for managing the acquisition of Aegis upgrades has four critical components. First, the model periodically distributes capability upgrades to both new and in-service ships using concurrent development and sequential integration and testing (I&T). Second, the model improves the efficiency of weapon system development and support by using modern software engineering processes that enable continuous development rather than the sequential process used under the legacy business model. Third, the model fosters competition by allowing the Navy to seek bids from multiple commercial vendors for developing individual components of the weapon system software. Finally, the model allows the Navy to leverage points of overlap in capability development across weapon systems. For example, each weapon system has a software component that manages detected threat tracks (a so-called "track manager"). Under the legacy business model, track managers were developed and implemented separately, but under the IWS business model, a single track manager would be available to all systems.

The OA character of the IWS business model promises substantial benefits[3] by allowing improvements to propagate across the Aegis fleet, introducing enhancements more quickly, and providing greater computing capabilities. However, moving from the legacy business model to the OA-based IWS model while maintaining a demanding operational schedule is challenging. The software and hardware upgrades to support the IWS business model must be installed across the entire Aegis fleet. The Navy is modernizing only three to four ships per year, with each ship upgrade requiring between 48 and 52 weeks. Thus, the Navy must maintain its legacy AWS for over 20 more years.[4] Further, the Missile Defense Agency's ballistic missile defense (BMD) program will have to find a place in the hardware and software schedules dictated by the OA plan. Finally, OA requires its own development, integration, and testing, which must occur at a faster pace than the historical norm for the Aegis program. Taken together, these factors make for complicated development, integration, and fielding activities.

---

[3] OA enables software components to work across a range of commercial computing hardware and interoperate with other software components.

[4] By law, ships within five years of their decommissioning date do not receive upgrades. All ships in the fleet after 2036 will be upgraded to the OA ACS.

Research Approach and Limitations

We used a multi-pronged approach to address our study questions. First, we conducted semistructured interviews with industry and government representatives from the Aegis enterprise, including the PEO IWS, Lockheed Martin, the Aegis Technical Representative (Aegis TECHREP), the Naval Surface Warfare Center (NSWC) Dahlgren Division, the NSWC Port Hueneme Division, the NSWC Corona Division, the Surface Combat Systems Center (SCSC), and the Combat Systems Engineering Development Site (CSEDS). These interviews focused on characterizing the legacy approach to developing, fielding, and supporting the AWS and on understanding each representative's view of how the IWS business model might affect the enterprise.

Second, we interviewed industry and government representatives from the Acoustic Rapid COTS Insertion (ARCI) and Ship Self-Defense System (SSDS) enterprises, including Raytheon and PEO Submarines. These interviews focused on understanding lessons learned from ARCI's and SSDS's unique experiences in transitioning to an OA-based approach.

Third, we collected historical workforce and facility usage data from key organizations and facilities in the Aegis enterprise. These data allowed us to characterize the historical effort involved in developing, integrating, and testing legacy baselines and ACBs and provided a basis for characterizing the choices and trade-offs involved in transitioning to the IWS business model.

Fourth, we developed a simulation model to estimate the effect of both the IWS business model and the Aegis modernization rate on the fleet. The simulation model allows the rate of software and hardware upgrades to vary independently of each other. In the context of this report, *drumbeat* refers to the periodicity of an update. For example, a software update drumbeat of two years means that PEO Integrated Warfare Systems develops and fields an AWS software upgrade every two years. Additionally, the model allows individual ships to receive either every upgrade or every other upgrade.

Finally, we developed a spreadsheet model to estimate the technical infrastructure required to develop, integrate, and test AWS upgrades. Using Naval Surface Warfare Center (NSWC) and prime contractor data on personnel, facility usage, and cost, we applied the model under varying assumptions regarding upgrade drumbeats and level of effort.

Our analysis of implications of the Navy's plan requires us to make various assumptions about, for example, the stability of funding for Aegis, shipbuilding plans and schedules, and ship availabilities for weapon system modernization and upgrades, among other issues discussed below. We based these assumptions on the most current Navy plan for modernization, shipbuilding, and upgrades at the time of our writing. In reality, however, future funding is unknowable, ship availabilities change routinely, and the expense of upgrades depends on countless factors outside the scope of our analysis. While we do not expect our findings to depend on minor changes to these parameters, we discuss potential risks below.

This report focuses on the development, integration, testing, and fielding of periodic updates to the Aegis fleet under the proposed IWS business model. Decisions made by the Navy in implementing the model will strongly affect Aegis training resources. Training resources—including instructors, equipment, and laboratory space—are limited and could constrain implementation. However, this report does not assess the impact of the IWS business model on Aegis training resources.

How the Legacy Business Model Differs from the IWS Business Model

The legacy and IWS business models differ substantially. Since its inception, the Aegis program has fielded a new version of the baseline system every five to six years. Under the legacy business model, a ship receives an initial AWS baseline and, potentially, an updated version at the midlife upgrade.[5] The IWS business model, by contrast, allows soft-

[5] Many surface combatants enter the shipyard at the midpoint in their expected service life to receive updates to their installed hull, mechanical, electrical, and combat systems.

ware and hardware improvements to be introduced on any modernized Aegis ship (i.e., one that has had both ACBs and TIs). Under this plan, a new ACB and TI are developed every four years. Individual Aegis ships receive every other upgrade. The IWS plan substantially alters the performance characteristics of the fleet. Once the AWS reaches a steady state (in approximately 2028), it will take about 7.5 years to install a given software upgrade across the entire fleet. Table S.1 compares the fleet attributes under the IWS and legacy business models.

The two business models also present different cost implications. As mentioned, the legacy business model involves installing an initial capability during construction, with one subsequent upgrade. Since individual ships receive no further upgrades, they incur no further fielding costs, and the capability remains as it is. Under the IWS business model, ship software and hardware is continually upgraded. Each upgrade incurs cost, both for the hardware itself and for the team required to install the upgrade.

However, the IWS business model also produces some cost efficiencies. Under the legacy model, each upgrade is installed on about 21 percent of the fleet; under the IWS business model, each upgrade reaches 96 percent of the modernized fleet. Essentially, development costs are spread across four times as many ships. Also, systems that depend on commercial hardware and software tend to have significantly lower initial installation costs.

But the IWS business model carries with it several sources of risk. The most basic risk stems from the fact that the plan differs fundamentally from the legacy business model in how it develops and fields capa-

Table S.1
Estimated Fleet Attributes Under the Legacy and IWS Business Models

Model	Software Upgrades per Year (average)	Hardware Upgrades per Year (average)	Hardware-Software Combinations in Fleet	Software Age (years)		Hardware Age (years)	
				Max.	Avg.	Max.	Avg.
Legacy			4.5	25.6	14.0	25.6	14.0
IWS	18.7	9.3	4.0	8.7	6.2	12.7	8.0

NOTE: Legacy data not available for blank cells.

bility upgrades. The Navy's use of the ARCI program for its submarine combat systems and the SSDS for aircraft carriers provides some related institutional experience and lessons learned, but Aegis differs from those programs in critical ways. Thus, the Navy should expect a uniquely complex fielding experience, proceed slowly when implementing this plan, and be prepared to derive its own lessons learned as it fields periodic upgrades to modernized ships and develops these upgrades from a CSL.[6]

Other sources of risk include the fact that multiple government stakeholders may have a vested interest in the legacy business model; the complexity of managing a CSL; the possibility that the Navy and the BMD program will compete for a limited pool of technical personnel, facility time, and access to the CSL; and the diversion of resources to the CSL from direct capability improvements.

The Navy can mitigate some of these risks by making capital investments in the CSL and software componentization, delaying investments in product-line development until the transition to the CSL is successful, streamlining government involvement in I&T to reduce schedule risk, enforcing requirements discipline, staggering TIs and ACBs, and harvesting lessons learned from ARCI and SSDS.

Effects of the IWS Business Model on Aegis Modernization and Fielding Rates

Generally speaking, the IWS business model improves Aegis fleet capabilities by spreading individual upgrades to all or parts of the fleet. That said, the Navy's PEO for Integrated Warfare Systems, working with the fleet and the Office of the Chief of Naval Operations, can make policy choices that affect both the infrastructure required for Aegis development and the capabilities delivered to the fleet. It can choose the rate (referred to in this report as a "drumbeat") at which periodic

6 The CSL is a master library that stores the code for all the Aegis applications and allows the Navy to develop several software components concurrently that can then be propagated to the fleet.

software and hardware upgrades occur. It can also decide whether ships will receive every upgrade or every other upgrade. Additionally, it can choose to field ACBs and TIs simultaneously or to stagger them. We developed a model to assess the effects of these various decisions.

Drumbeat Decisions

We explored the effects of two-, four-, and six-year drumbeats for ACB and TI insertions. Additionally, we analyzed the effects of giving ships every upgrade versus every other upgrade. Table S.2 compares the effects of the drumbeats examined for the IWS business model with those of legacy practices in terms of average and maximum age of software and hardware.

Under the IWS business model, more of the fleet has newer software and hardware. The average age of these components under the legacy business model is 14 years, and the maximum age is almost 26 years. With a drumbeat of six-year insertions under the IWS business model, these ages drop to just under seven years and almost nine years, respectively. This also means that ACBs and TIs do not stay in the fleet as long as they do under the legacy business model. Thus, there are fewer Aegis versions in the fleet to support under the IWS model. Under the legacy business model, an average of 4.5 Aegis versions are deployed in the fleet at a given time. Meanwhile, under IWS model, an average of only two are deployed. Thus, IWS lowers the average age of the technology present in the fleet and brings that technology closer to the industry's current hardware obsolescence cycle.

If the drumbeat quickens to four-year insertions, the maximum age declines from nine years to six, and the average age declines from

Table S.2
Effects of Different Drumbeats on the Average and Maximum Ages of Hardware and Software

Age of Hardware and Software (years)	Legacy Business Model	IWS Business Model		
		6-Year	4-Year	2-Year
Average	14	7	5	< 3
Maximum	26	9	6	> 3

seven to five. If the drumbeat quickens even further—to two-year insertions—the maximum age drops to just over three years and the average age to just under three years. However, speeding up the insertions means that more ships are upgraded each year. At a drumbeat of six years, 11 ships per year receive upgrades, but with a drumbeat of two years, that number climbs to 34. This increase has important implications for the Navy's ability to make sufficient ships available for upgrading and for training crews in the new concepts.

Upgrade Decisions

The difference in the effects of getting every upgrade versus every other one out to the fleet is also significant. For example, when the ACB drumbeat is two years but a ship receives only every other upgrade, the average age of the software increases from three to four years, and the maximum age increases from just over three to just over five years. Also, the average number of hardware-software combinations deployed in the fleet rises from 2.6 to 5.5.

The current IWS business model calls for four-year software and hardware upgrades, with ships receiving every software upgrade and every other hardware upgrade. This results in individual upgrades being installed on about 43 percent of the fleet (better than legacy business model results of 21 percent), but the process would be about half as efficient if ships were to get every upgrade. The model calls for installing hardware on eight ships and software on 17 ships per year. This actually increases the number of hardware-software combinations in the fleet, which will likely increase in-service support costs and interoperability issues.

Implications for Development

As the Navy considers alternative upgrade intervals, it must balance upgrade size and frequency. Smaller, more frequent upgrades will distribute capability improvements across the fleet more quickly but will mean that the fixed costs of I&T will consume more of the fixed IWS budgets. Larger, less frequent updates will distribute capability more slowly but will result in the fixed costs of I&T consuming less of the fixed IWS budgets.

The choice of TI intervals requires the additional consideration of coordinating computing and networking hardware upgrades with the industry that produces the COTS equipment. ARCI has upgraded its hardware roughly every two years, which allows it to minimize procurement and in-service costs while leveraging recent, if not state-of-the-art, COTS equipment. Integrating new hardware every two years would be especially challenging for Aegis. However, the Navy may be able to mitigate the in-service cost of slower drumbeats by warehousing retired computing hardware and using the parts as spares. To date, Aegis has not needed the computing capacity that would be provided by more frequent upgrades.

Recommended Modernization Rate

We agree with the current IWS plan to field ACB and TI upgrades on a four-year drumbeat. In our proposed implementation approach, every ACB and TI upgrade is installed on every Aegis ship over the four-year period. Further, the ACB and TI upgrades are offset by two years. Figure S.1 illustrates this proposed approach. In addition to new computer hardware, TI upgrades include software fixes in response to computer program change requests (CPCRs) identified during the preceding ACB, as well as modifications to the AWS required to support ACS upgrades. For example, TI-18 would include software fixes identified by the fleet operating with the ACB-16 upgrades. Table S.3 shows the fleet attributes under three models: legacy, IWS, and RAND.

Why Four Years for ACB?

We recommend a four-year ACB drumbeat to balance the desire to deploy new capabilities with the risk of compressed I&T times and the disruption of each ship's operations. Aegis historical development indicates that a faster drumbeat would be difficult to execute. Furthermore, a faster rate would devote a prohibitively large fraction of Aegis technical resources to I&T and constrain development efforts critical to providing mature technology for subsequent ACBs. Finally,

Figure S.1
RAND's Proposed ACB/TI Implementation

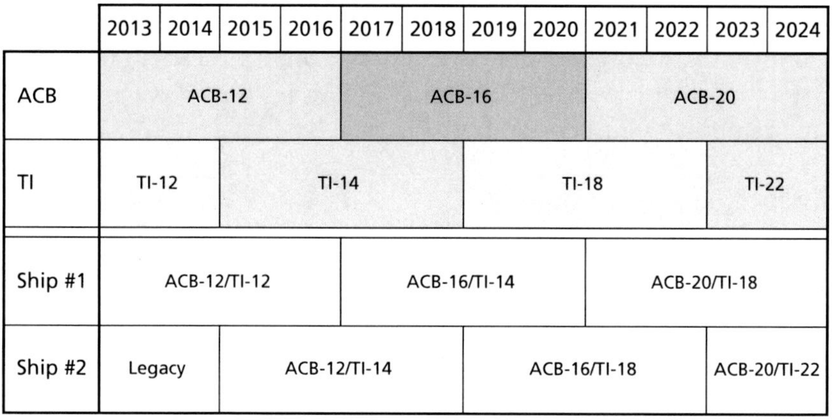

RAND *RR161-S.1*

the planned capabilities in the Aegis technology roadmap fit easily into four-year cycles.

Why Four Years for TI?

The I&T burden for TI is considerably lower than for ACB. The TI drumbeat must support the computing power required by the ACB capabilities. Normally, this would suggest a faster TI drumbeat than ACB drumbeat, but the current suite of ACB upgrades does not require all of the additional computing power offered by the switch to com-

Table S.3
Fleet Attributes Under Three Plans

Model	Software Upgrades per Year (average)	Hardware Upgrades per Year (average)	Software-Hardware Combinations in Fleet	Software Age (years)		Hardware Age (years)	
				Max.	Avg.	Max.	Avg.
Legacy			4.5	25.6	14.0	25.6	14.0
IWS	18.7	9.3	4.0	8.7	6.2	12.7	8.0
RAND	18.7	18.7	3.0	8.7	6.2	8.3	6.2

NOTE: Legacy data not available for blank cells.

mercial hardware. Furthermore, installing new commercial hardware on the required number of ships under the IWS business model is expensive. A four-year drumbeat minimizes the potential risks inherent in deviating from the commercial cycle. Also, including software fixes in each hardware upgrade increases opportunities to improve the stability of the Aegis code, respond to issues identified by operators, and improve training stability.

Why Stagger Insertions?

Offsetting the four-year ACB and TI cycles balances deployed capabilities and development risk and offers three advantages. First, it isolates software and hardware development efforts from one another. Installing upgraded software on mature hardware enables the rapid identification of issues in either the hardware or software. Second, this approach incorporates software fixes in both the software and hardware upgrades, doubling the opportunities to incorporate such fixes and support ACS element upgrades. Third, this approach allows the Navy to level-load the demand on the Aegis technical infrastructure.

Acknowledgments

This research could not have been accomplished without the assistance of many individuals. Chris Deegan, deputy program executive officer for Integrated Warfare Systems, encouraged and supported this research effort. Bill Bray, Myron Liszniansky, Kathy Emery, and LCDR Joel MacRitchie, also at the Navy's PEO for Integrated Warfare Systems, graciously shared their time and expertise.

Numerous individuals in the Dahlgren and Port Hueneme divisions of the NSWC shared their knowledge of ACS engineering processes and infrastructure. In particular, we would like to thank Brian Seay at Dahlgren and Michael Horton at Port Hueneme for sharing their extensive knowledge of their organizations' roles in Aegis technical support. Kevin Kolb and the personnel at Aegis TECHREP provided valuable insights into the Aegis development process.

The entire engineering team at Lockheed Martin Maritime Sensors and Systems supported our research effort with data, technical expertise, and advice. The White House Diner in Moorestown, New Jersey, always sustained our efforts.

Reuben Pitts, former head of the Warfare Systems Department at the NSWC Dahlgren Division, and Ken Munson at RAND offered valuable insights and suggestions on earlier drafts of the report that greatly improved the presentation and research. RADM (ret.) Kate Paige fortified the study team with her in-depth knowledge of Aegis and missile defense. Roland Yardley shared his expertise and knowledge of the Aegis system, especially within the training enterprise.

While these individuals, and others too numerous to mention, provided information and comments during our research effort that had a positive impact on the resulting report, we are solely responsible for the interpretation of the information and data and the conclusions drawn.

Abbreviations

ACB	advanced capability build
ACS	Aegis combat system
AMOD	Aegis modernization
APB	advanced processing build
ARCI	Acoustic Rapid COTS Insertion
AWS	Aegis weapon system
BL	baseline
BMD	ballistic missile defense
CAL	common asset library
CG	U.S. Navy Hull Classification System designation for a guided missile cruiser
COTS	commercial off-the-shelf
CPCR	computer program change request
CSEDS	Combat Systems Engineering Development Site
CSL	common source library
DDG	U.S. Navy Hull Classification System designation for a guided missile destroyer
ESLOC	equivalent source lines of code
FTE	full-time equivalent
FY	fiscal year
I&T	integration and testing

IAMD	Integrated Air and Missile Defense
IWS	integrated warfare systems
IWSL	Integrated Warfare Systems Laboratory
MDA	Missile Defense Agency
MILSPEC	military specification
MS2	Maritime Systems and Sensors
MY	man-year
NSCC	Naval Systems Computing Center
NSWC	Naval Surface Warfare Center
OA	open architecture
PEO	Program Executive Office
SCSC	Surface Combat Systems Center
SSBN	attack ballistic missile
SSDS	Ship Self-Defense System
SSL	single source library
SSN	attack submarine
TECHREP	Technical Representative
TI	technology insertion
TPR	test program review

CHAPTER ONE

Introduction

The Navy's transition from its legacy Aegis business model to its new Integrated Warfare Systems (IWS) business model[1] may introduce new challenges and risks for the fleet and for the enterprise that develops and fields the Aegis weapon system (AWS). Under the legacy business model, the AWS used proprietary software operating on military-specification (MILSPEC) computing hardware. Upgrades to the Aegis combat system (ACS) were developed every five to six years and fielded only to new-construction ships and those receiving a midlife upgrade.[2] Older baselines were upgraded to support additional capabilities, fix computer software errors, and support upgrades to ACS elements. Upgrades or modifications to deployed Aegis systems to support ACS element upgrades put a significant demand on the Aegis technical infrastructure. The new IWS business model will use open-architecture (OA) software operating on commercial off-the-shelf (COTS) computing hardware. The IWS model will also involve periodic upgrades to all ships, both new and in-service. Software will be upgraded through advanced capability builds (ACBs) every four years. These upgrades will occur independently of computing hardware

[1] The IWS business model is articulated in the Program Executive Office (PEO) Integrated Warfare Systems Acquisition Management Plan (2013).

[2] AWS refers specifically to the computer software and hardware, radar system (SPY-1), and vertical launch system onboard an Aegis ship. The additional sensors, communication systems, weapons, and countermeasures are part of the broader ACS.

upgrades, called technology insertions (TIs), which will take place every four years, with individual ships receiving every other upgrade.[3]

The IWS business model for managing the acquisition of AWS upgrades has four critical components. First, the model periodically distributes capability upgrades to both new and in-service ships using concurrent development and sequential integration and testing (I&T). Second, the IWS business model aims to improve the efficiency of weapon system development and support by using modern software engineering processes that enable continuous development rather than the sequential process inherent under the legacy business model. Third, the IWS business model attempts to foster competition by allowing the Navy to seek bids from multiple commercial vendors for developing individual components of the weapon system software. Finally, the model ideally allows the Navy to leverage points of overlap in capability development across weapon systems. For example, each weapon system has a software component that manages detected threat tracks (a so-called track manager). Under the legacy business model, track managers were developed and implemented separately, but under the IWS business model, the Navy intends to develop a single track manager that would be available to all systems.

The IWS business model pertains primarily to a development program (see PEO for Integrated Warfare Systems, 2013). However, this business model will affect the entire Aegis lifecycle. The development, integration, and testing schedule will quicken to support a four-year cycle time. The Navy will have to support multiple ship upgrades each year. The in-service support infrastructure will no longer have to maintain MILSPEC software and hardware for the life of the ship; rather, it will maintain a constantly evolving set of COTS-based computing hardware and middleware. In this report, we focus on the development, integration and testing, and fielding of Aegis upgrades. Specifically, the report attempts to answer the following questions:

---

[3] Individual ACBs and TIs are named according to the year of their fielding, so ACB-08 is the name of the ACB schedule for fielding in 2008.

- How does the Navy currently develop, test, and field upgrades to the AWS, and how will that process change under the IWS business model?
- How does the IWS business model affect AWS modernization and fielding rates in terms of both the technical infrastructure and fleet capabilities?
- What modernization rate under the IWS business model should be recommended to the Navy to balance fleet capability, risk, and cost?

It is important for the Navy to answer these questions in a timely manner. The Navy's surface fleet has already begun to transition to an OA construct operating on COTS computer equipment. Without a well-thought-out modernization program, the fleet will experience increasingly challenging obsolescence issues. Additionally, the introduction of new capabilities into the Aegis fleet is likely to quicken over the next decade due to ballistic and cruise missile defense requirements. The Aegis fleet is the backbone of the Navy's surface fleet and, with these new capabilities, it will remain so for decades.

Research Approach

In the first decade of the 21st century, the Navy's PEO for Integrated Warfare Systems fielded four configurations of the AWS. This report examines the technical infrastructure required to develop future versions of OA Aegis upgrades.

First, we conducted semistructured interviews with industry and government representatives from the Aegis enterprise, including PEO Integrated Warfare Systems, Lockheed Martin, the Aegis Technical Representative (Aegis TECHREP), the Naval Surface Warfare Center (NSWC) Dahlgren Division, the NSWC Port Hueneme Division, the NSWC Corona Division, the Surface Combat Systems Center (SCSC), and the Combat Systems Engineering Development Site (CSEDS). These interviews focused on characterizing the legacy approach to developing, fielding, and supporting the AWS and on understanding each representative's view of how the IWS business model might affect the enterprise.

Second, we interviewed industry and government representatives from the Acoustic Rapid COTS Insertion (ARCI) and Ship Self-Defense System (SSDS) enterprises, including Raytheon and PEO Submarines. These interviews focused on understanding lessons learned from ARCI's and SSDS's unique experiences in transitioning to an OA-based approach.

Third, we collected historical workforce and facility usage data from key organizations and facilities in the Aegis enterprise. These data allowed us to characterize the historical effort involved in developing, integrating, and testing legacy baselines and ACBs and provided a basis for characterizing the choices and trade-offs involved in transitioning to the IWS business model.

Fourth, we developed a simulation model to estimate the effect of both the IWS business model and the Aegis modernization rate on the fleet. The simulation model allows the drumbeat of software and hardware upgrades to vary independently of each other. In the context of this report, *drumbeat* refers to the periodicity of an update. For example, a software update drumbeat of two years means that PEO Integrated Warfare Systems develops and fields an AWS software upgrade every two years. Additionally, the simulation model allows individual ships to receive either every upgrade or every other upgrade.

Finally, we developed a spreadsheet model to estimate the technical infrastructure required to develop, integrate, and test AWS upgrades. Using Naval Surface Warfare Center (NSWC) and prime contractor data on personnel, facility usage, and cost, we applied the model to varying assumptions regarding upgrade drumbeats and level of effort.

This report focuses on the development, integration, testing, and fielding of periodic updates to the Aegis fleet under the proposed IWS business model. Decisions made by the Navy in implementing the model will strongly affect Aegis training resources. Training resources—including instructors, equipment and laboratory space—are limited and could be a constraint during implementation. This report, however, does not assess the impact of the IWS business model on Aegis training resources.

A previous report documented the methods and findings of this research effort but incorporated proprietary information. This report does not contain any proprietary information and incorporates the most recent Navy Aegis modernization approach.

Organization of This Report

Chapter Two describes the IWS business model and the Aegis fleet's transition to an OA-based approach. Chapter Three describes the scope of the Navy's Aegis technical enterprise, as well as addresses the organizations that participate in deploying and maintaining the Aegis fleet and examines the nature of their participation. Chapter Four describes the impact of Aegis modernization rates and PEO Integrated Warfare Systems decisionmaking on the Aegis fleet. Chapter Five discusses the implications of that decisionmaking for the Aegis development enterprise. Chapter Six explains the risks that PEO Integrated Warfare Systems will face as it implements its business model. Chapter Seven examines the lessons learned from ARCI and SSDS as they apply to the AWS. Chapter Eight presents our proposed implementation of AWS upgrades and summarizes our analysis.

CHAPTER TWO

The IWS Business Model for Aegis Acquisition

In this chapter, we describe the IWS business model and the choices that must be made over the course of its implementation. We distinguish what we see as the implied objectives of the business model (the "ends") from the investments that are being made to execute the model (the "means").

Plan and Objectives

The IWS business model involves five fundamentally distinct objectives. We consider each in turn and relate them to the legacy approach to acquiring weapon systems.

Distribute Periodic Capability Upgrades to New and In-Service Ships

Under the legacy business model for acquiring weapon system upgrades, the Navy developed capabilities for new-construction ships and upgraded in-service ships at midlife. Each upgrade, historically called a *baseline* (BL), was composed of both computing hardware and software. Development timelines were set by new construction fielding schedules, and weapon system fielding schedules could slip with delays in development. Upgrades were made occasionally to individual AWS BLs to support relatively minor capability enhancements and often to support ACS element upgrades.

Under the IWS business model, all ships—both new and in-service—will receive upgrades on a regimented and periodic basis. Software upgrades, or ACBs, will be fielded every four years, concurrently with computing hardware upgrades, or TIs, also fielded every four years.[1] ACBs and TIs will be fielded on new-construction and modernized in-service ships during planned nine-week ship availabilities. Ships will receive every ACB upgrade and every other TI upgrade. The fielding schedule will be set independently of both the development and construction schedules, and it will be regimented in the sense that ACBs and TIs will be conducted regardless of whether development meets or misses milestones; each ACB will harvest sufficiently mature technology to be integrated and tested in time to meet the fielding schedule.

Each ACB or TI involves phases of planning, I&T, and fielding. During the planning phase, the composition of ACBs is defined. The capabilities incorporated in an individual ACB may include full-fledged programs of record, activities designed to address specific concerns raised by the fleet, software maintenance efforts, or any combination of the three. Incorporation of individual capabilities is constrained by the requirement that a technology be sufficiently mature for integration and testing within the timeframe necessary to meet the fielding schedule. Obviously, the incorporation of capabilities in an ACB or TI is also subject to budget availability. The I&T phase integrates capabilities into the broader combat system, tests the Aegis system in live and simulated environments, and certifies it for deployment. Finally, in the fielding phase, a certified capability is distributed to the fleet.

A BL number designates the Aegis combat system configurations fielded to individual ships. For example, the common source library (CSL)[2] following the completion of ACB-12 development is used to field Baseline 9 combat system configurations. The specific BL 9 configurations include the following:

- BL 9A: Air Defense Cruisers (CGs 59–64)

---

[1] As mentioned in Chapter One, individual ACBs and TIs are named according to the year of their fielding, so ACB-14 is the name of the ACB scheduled for fielding in 2014.

[2] A CSL is a shared software library that is continuously updated whenever problems are discovered in the fleet so that fixes or improvements can be made once and then propagated to other future and in-service ACBs.

- BL 9C: Integrated Air and Missile Defense (IAMD) DDG (DDGs 51–112)
- BL 9D: New-construction IAMD DDG (DDG 113 and follow-ons)
- BL 9E: Aegis Ashore with ballistic missile defense (BMD) only.

The regimented and periodic upgrade time is perhaps the defining feature of the IWS business model. Figure 2.1 depicts a notional timeline under the IWS business model.

Improve Efficiency of Weapon System Development and Support

Under the legacy business model, the Navy develops and supports capability upgrades for a given system as separate baselines. New base-

Figure 2.1
Notional Development Timeline Under the IWS Business Model

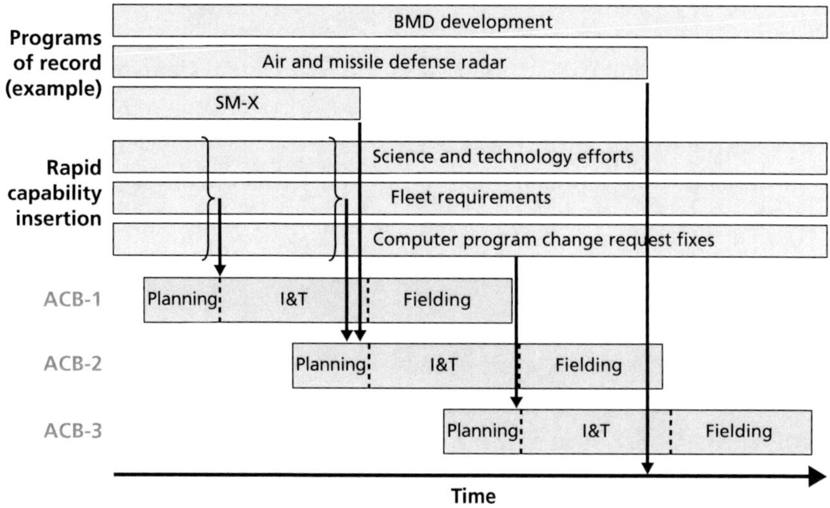

NOTE: BMD refers to the Missile Defense Agency's (MDA's) BMD program. SM-X = a given standard missile derivative.

RAND *RR161-2.1*

line development initiatives begin by "cloning"[3] the software of a previous baseline. After the baseline is certified, it is maintained separately. Navy officials sometimes refer to this approach as "clone and own." One of the implications of this approach is that fixes or improvements to a given baseline do not propagate across the fleet unless the repair is funded for each baseline to which it applies.

Under the IWS business model, however, weapon system software undergoes a continuous development process, facilitated by a CSL. The management of this CSL, however, is one of the most significant sources of risk associated with the IWS model, as we will discuss in greater depth later.

Promote Competition in Weapon System Development

Under the legacy approach, the Navy selects a prime contractor to maintain responsibility for essentially all aspects of weapon system development. The Navy has an opportunity to open development to competition, but only at the system level.

Under the IWS approach, the Navy intends to solicit separate bids for individual components of the weapon system software. For example, it may issue separate requests for the design of radar processing algorithms, display systems, or fire-control systems. A prime integrator will then integrate components that may have been developed separately. Our discussions with Navy officials suggest that such component-level competition is anticipated to foster innovation and potentially reduce cost.

Leverage Capability Development Across Weapon Systems

The Navy develops and maintains multiple combat systems, including Aegis, SSDS, the Littoral Combat Ship system, and DDG-1000 (*Zumwalt*-class destroyer). Each system is designed to fill a unique need on a specific platform, but the basic functionality of the weapon sys-

---

[3] Cloning refers to the practice of beginning the development of a new baseline with the software from the current production baseline. Upgrades are implemented to this cloned software and form the basis for the new baseline. Changes made to the cloned software are generally not integrated into "parent" baselines or other clones.

tems' software overlaps. For example, each weapon system maintains a software component that manages detected threat tracks (a so-called track manager). Under the legacy approach, each weapon system develops and maintains its own track manager component—and all other components, for that matter.

Under the IWS business model, common software components, such as the track manager, would be developed only once and made available for use by all of the Navy's weapon systems. Moreover, when problems (e.g., software bugs) are discovered in any particular component, repairs can be made and propagated throughout the Navy's suite of weapon systems. Common components will be developed and shared across systems through a common asset library (CAL).[4]

Integrate Aegis and the Missile Defense Agency's Ballistic Missile Defense Program

Under the legacy business model, the BMD program and Aegis are developed separately. Under the IWS business model, they will be developed jointly, which is to say they will share the same software suite and hardware components. For example, ACB-12/BL 9 will involve installing a multi-mission signal processor that enables the ballistic missile and air defense modes of the AWS to run concurrently for an IAMD capability.[5] In addition, ACB-12 combines the software in a single suite. Further improvements to anti-air warfare and missile defense will be made through the ACB/TI process. This will require the Navy and the Missile Defense Agency (MDA) to coordinate during the planning phases of each ACB to ensure that the combined updates do not exceed the restrictive I&T timelines of the IWS business model.

Note that, as a result of this effort, all Aegis DDGs that complete the Aegis Modernization Program will be IAMD capable.[6] Over time, this will significantly increase the size of the BMD-capable fleet. Fur-

---

4 A CAL is shared across combat systems, whereas a CSL is specific to a single combat system. Some, not all, components of a combat system's CSL are part of the CAL.

5 The multi-mission signal processor (MMSP) will not be installed on any Aegis Cruisers.

6 CGs 52–58, which received ACB-08, will not be BMD-capable.

ther, improvements to BMD functions incorporated into individual ACBs will propagate quickly throughout the Aegis fleet.

Table 2.1 summarizes some of the features that distinguish the IWS business model from the legacy approach to acquiring weapon systems. Core Aegis refers to non-BMD IAMD development efforts for U.S. surface combatants. Total Aegis efforts include the core Aegis program, efforts related to BMD, and development related to international activity.

Enabling Investments

The IWS business model's objectives are enabled by a variety of investments that the Navy has made and continues to make in the AWS. These investments can be quite significant in terms of cost and time—so significant, in fact, that they might be misconstrued in offi-

Table 2.1
Comparison of the Legacy and IWS Business Models

Dimension	Legacy	IWS
Development process	Sequential development ("clone and own")	Concurrent development with CSL
	Integration and testing occurs after development passes predefined milestones	Integration and testing occurs on a regimented and periodic basis (drumbeat)
	BMD and Aegis developed as essentially different systems	Integrated air and missile defense
Roles and responsibilities	All development managed by prime contractor	Integration and testing managed by prime contractor; development handled on competitive basis
Acquisition strategy	Development fielded to new-construction ships	Development fielded to both new-construction and in-service ships
	Separate development for different combat systems	Product line approach with CAL

cial U.S. Navy documents as ends rather than the means necessary to achieve the Navy's broader objectives. In this section, we consider these investments in turn.

First, the Navy has decoupled the AWS from the ship. This means that the same combat systems can be deployed across multiple ship classes—for example, across different variants of cruisers, destroyers, and as-yet-unplanned future ships. This decoupling has been achieved by designing the physical architecture of the combat system in a way that ensures broad applicability of the physical plant and by modernizing the in-service fleet to host updated Aegis weapon systems. This investment contributes to the objective of distributing combat system capability across a broad set of ships by relaxing necessary ship conditions.

Second, the Navy has transitioned from MILSPEC computing hardware to COTS technology. Historically, the AWS has relied on customized computing hardware (e.g., AN/UYK-7, AN/UYK-43). However, beginning with BL 6.3, the Navy began the transition to COTS processors. In theory, COTS hardware allows the U.S. Navy to leverage Moore's law–type advances in processing power and to avoid the prohibitive cost of developing and supporting its own processors. Thus, COTS hardware helps increase capability and makes development more efficient.

Third, the Navy has decoupled weapon system computing hardware and software through the use of middleware. Middleware is software that interfaces between computing hardware and the weapon system's operating system and applications. In theory, robust middleware with standardized interfaces would allow computing hardware to be changed without also changing the weapon system software and vice versa. Middleware allows the Navy to develop capability upgrades for a broader portion of the Aegis fleet (i.e., ships with a range of TIs could receive ACB upgrades) and allows software development to proceed in a way that is less sensitive to hardware specifications.

Fourth, the Navy is investing in a modular software architecture with published government-owned and authenticated interfaces. Software architecture defines the basic components of a software system, the relationships between them, and ways in which they collectively relate to system capabilities (Pfleeger and Atlee, 2005). A modular software architecture is one in which there is no overlap in functionality across software components and in which the interfaces between components are well defined. A modular architecture can reduce development cost by isolating problems within components and minimizing the impact of local problems on broader system functionality. Publishing the interfaces (i.e., documenting interface descriptions so they can be shared as needed) is a prerequisite for allowing broader competition.

Finally, the Navy is investing in software development processes and infrastructure that facilitate software reuse. As mentioned earlier, it is developing a CSL in ACB-12/BL 9 to allow concurrent development across ACBs. The CSL is a critical component of the IWS business model that is necessary for concurrent development and for distributing periodic capability upgrades to the Aegis fleet. Separately, the Navy plans to develop a CAL through which mature components can be shared across weapon systems. The Joint Track Manager will be the first addition to the CAL. Although the CAL will support the objective of leveraging capabilities across weapon systems, it will not directly affect the objective of distributing capability upgrades.

It is important to clarify the role of ACB-12/BL 9 in the transition to the IWS business model. As a result of the investments made in developing ACB-12, PEO Integrated Warfare Systems will have a CSL for the AWS that will support subsequent ACB and TI efforts. ACB-12 is not necessarily representative of future ACB upgrades. It is an extremely large effort that would not fit within the timelines envisioned by the IWS model, but it is necessary for implementing the correct software architecture.

Although the CSL and CAL both support software reuse, they support separate objectives of the IWS business model. In fact, the Navy's choices to implement the CAL and the CSL are fundamentally independent: One does not come hand in hand with the other. As discussed in later chapters, the CSL and CAL are both sources of risk, and separating their implementation is one way to mitigate that risk.

Implementation Choices

PEO Integrated Warfare Systems has several choices to make when implementing its business model. These choices include

- the time between computing hardware upgrades (i.e., the TI interval)
- the time between combat system software upgrades (i.e., the ACB interval)
- the size and complexity of software upgrades (i.e., ACB size)
- the frequency with which individual ships receive hardware and software upgrades
- the pace at which in-service ships' weapon systems are modernized.

The IWS business model specifies four-year TI and ACB intervals but leaves the size of future ACBs, the frequency of upgrades, and the pace of modernization unspecified. This report analyzes alternative options for each of these choices and how they would affect the Navy's personnel, processes, and facilities. This range of alternatives is summarized in Table 2.2.

Table 2.2
Current and Alternative Options for Implementing the IWS Business Model

Choice	IWS Business Model	Range of Alternatives
TI interval	4 years	2–4 years
ACB interval	4 years	2–4 years
ACB size	Unspecified	Fix software bugs, simultaneously upgrade computing hardware, integrate weapon system components, and add combat system capability
Ship update frequency	Every other update	Every update or every other update
Modernization pace	1 or 2 DDGs per year	0–3 CGs and 0–3 DDGs per year

SOURCE: Information on the modernization pace is from O'Rourke, 2012.

Assumptions

The IWS business model does not fully specify development and fielding timelines, but conversations with Navy officials allowed us to fill in these gaps with several assumptions.

Our first assumption is that exactly one ACB-TI combination—the most recently certified such combination—is being fielded at any given time. This assumption was reflected in Figure 2.1, earlier in this chapter, by the fact that ACB fielding intervals are consecutive and nonoverlapping in time. We assume that the Navy will adopt a policy such that, when a given ship is upgraded, it receives the most recently certified ACB-TI combination. We view this assumption as uncontroversial because it is exactly in the spirit of distributing the most recent capability upgrades to the widest possible segment of the fleet; there is no reason that the Navy would field old or uncertified technology.

Our second assumption is that exactly one ACB-TI combination is in the I&T phase of development at a given time. This was reflected in Figure 2.1 by the fact that the I&T phases of subsequent ACBs were consecutive and nonoverlapping in time. A critical feature of the IWS business model is to integrate and test one ACB or TI at a time. Using a sequential I&T process, software bugs and subsequent fixes need to be funded only once. Parallel I&T processes, on the other hand, require software fixes to be paid for multiple times and increase the probability that the same software bug will remain in subsequent ACBs or TIs. Note that this assumption is not the same as saying there is only one ACB in development at a time. In fact, the IWS business model calls for concurrent development.

An important implication of these assumptions is that the next ACB is in I&T while the current ACB is being fielded. Additionally, the fielding interval determines the time available for I&T. For example, an ACB interval of four years implies that integration and testing must be completed in four years as well.

Observations

Several observations are in order. First, the Navy's weapon systems are ultimately developed to fight wars, yet the IWS business model specifies its objectives in terms of acquisition, not warfighting. Of course, there are virtuous reasons to pursue faster and more broadly distributed upgrades, more efficient development processes, competition, and software reuse, particularly when budgets are tight, as is the case today. To the extent that the model resembles ARCI, the fleet may infer a capability-based rationale for the model from the capability improvements the submarine community experienced as the result of ARCI. As we show later in this report, the fleet may bear some significant risks and costs under this model, and over time, it may become more important for the Navy to justify it more explicitly in terms of warfighting capability.

Also, the IWS business model specifies changes in the development process and thus has *direct* implications for the people and facilities involved in developing weapon systems for the surface fleet (the focus of this report). However, weapon systems, once developed, must also be fielded and supported, and the sailors who use the systems must be trained; ultimately, the systems are designed for combat. Thus, the model will certainly have *indirect* effects on Navy resources that are tasked to field and support weapon systems and to train sailors. For example, more frequent upgrades could impose an undue burden on sailors who must keep pace with capability upgrades through training. Less frequent upgrades, however, would cause ships to be deployed longer with the same software and hardware capabilities, which could increase the demand for in-service support. The model also entails potentially significant indirect effects for international partners who purchase weapon systems through foreign military sales. Figure 2.2 is an influence diagram illustrating both the direct and indirect effects of the IWS business model on the people, processes, and facilities involved with a weapon system throughout its lifecycle.

Finally, the Navy has already made significant investments toward the IWS business model that would make it difficult, or even impossible, to return to using the legacy approach. The transition to COTS

Figure 2.2
Influence of the IWS Business Model on Weapon System Lifecycle

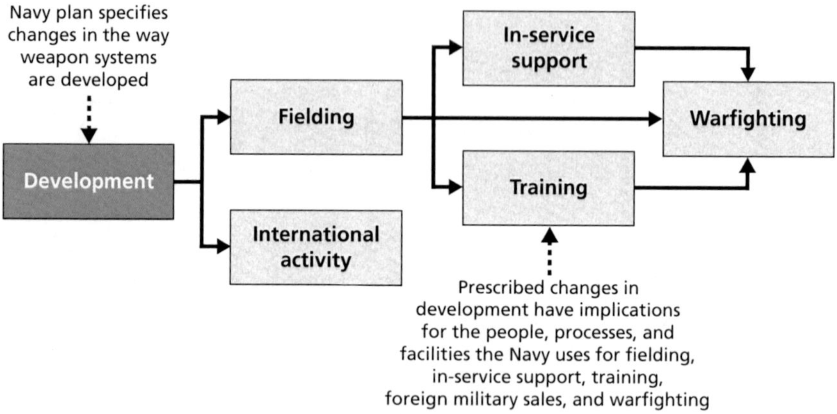

RAND *RR161-2.2*

hardware, for example, makes it infeasible to upgrade only at midlife, since commercial processor manufacturers do not support shipsets for that length of time. ACB-12 development is well under way and will result in both a usable CSL and the first addition to the CAL (the Joint Track Manager). Thus, notwithstanding implementation choices, the question facing the Navy is not if, but how and at what risk and cost, it will move forward with the IWS business model.

CHAPTER THREE

Aegis and the Aegis Enterprise

Introduction

A large enterprise of industry and government organizations and a network of development and test facilities support AWS. The IWS business model is expected to affect these organizations and facilities in a variety of ways, and identifying these impacts naturally requires an understanding of the baseline approach to AWS development. That is to say, one must first understand how the enterprise is presently organized and how the respective organizations and facilities contribute to developing the AWS. Thus, the purpose of this chapter is to answer the following questions:

- What does each facility and organization contribute to the Aegis enterprise?
- How is level of effort distributed across the enterprise in ways that may change under the Navy's plan?

Approach

Analysis of the AWS benefits from readily available documentation on the AWS, including system-engineering plans provided by PEO IWS. In contrast, no definitive U.S. Navy document describes the roles and responsibilities of the various organizations and facilities involved with AWS development. Moreover, our informal conversations with individuals across the Navy suggest that the enterprise is so vast that few individuals have an enterprise-wide view of exactly "who does what" with regard to Aegis. This is not surprising, given the size of the program and the fact that it has evolved considerably since its inception in the 1960s in response to changes in acquisition policy (e.g., Goldwater-Nichols), the maturing of system engineering best practices (e.g., calling for closer integration of government and industry, moving testing to an earlier point in the development process), and funding (e.g., the Reagan buildup, the end of the Cold War). Whatever the explanation, the point is that roles and responsibilities were not an input in our study and therefore had to be derived.

Thus, we took an empirical approach to establishing baseline roles and responsibilities and assessing levels of effort. First, we conducted semistructured interviews with representatives across the Aegis enterprise.[1] These interviews focused on understanding the current roles of the organizations in the enterprise, how each organization interacts with the others, and how individuals believe organizational roles might change under the IWS business model. This first step offered a qualitative perspective on the Aegis development enterprise. Second, we collected detailed historical data on facility usage and manpower. These data allowed us to break down facility usage, manpower, and cost over time by baseline and ACB, as well as by phase of the weapon system lifecycle. This provided a quantitative perspective on relative roles and responsibilities in the Aegis enterprise.[2]

---

1 Interviews were held with Lockheed Martin, including the Naval Systems Computing Center (NSCC) and Production Test Center (PTC); NSWC Dahlgren, including the Integrated Warfare Systems Laboratory (IWSL) and Aegis Training and Readiness Center; NSWC Port Hueneme; Aegis TECHREP, including CSEDS; SCSC at Wallops Island, Virginia; and NSWC Corona.

2 RAND collected historical manpower data from Lockheed Martin, NSWC Dahlgren, NSWC Port Hueneme Division, and Aegis TECHREP. Historical facility usage data collected from Lockheed Martin for CSEDS and NSCC; SCSC usage data came from permanent staff at SCSC; IWSL usage data came from NSWC Dahlgren. A more detailed analysis of proprietary workforce and facility usage is presented in the proprietary report.

Outline of Chapter

In this chapter, we examine the roles and responsibilities of key organizations and facilities in the Aegis enterprise before assessing level of effort across these entities.

Aegis Enterprise

Organizations

Many organizations have a role in Aegis development, integration, testing, fielding, and support. Six of the organizations we studied had some of the most important roles in the Aegis lifecycle. Their roles, as surmised from our interviews, are summarized in Table 3.1. Figure 3.1 shows the distribution of effort Lockheed Martin, NSWC Dahlgren, and NSWC Port Hueneme expend across phases of the Aegis lifecycle.[3]

Table 3.1
Roles of Organizations Across Aegis Lifecycle as Surmised by Interviews

	Phase in Weapon System Lifecycle					
Organization	Development	I&T	Fielding	In-Service Support	Training	Other
PEO IWS	√	√	√	√	√	√
Lockheed Martin	√	√				
Aegis TECHREP	√	√				
NSWC Dahlgren	√	√	√	√	√	√
NSWC Port Hueneme	√	√	√	√	√	
SCSC		√		√	√	
NSWC Corona		√				√

[3] In our analyses of workforce and facility usage time series, we associate the primary thrust of baseline development with activity occurring prior to test program review (TPR), the primary thrust of integration and testing with activity occurring between TPR and weapon system certification, and the primary thrust of operations and support with activity occurring after weapon system certification.

Figure 3.1
Distribution of Lockheed Martin, NSWC Dahlgren, and NSWC Port Hueneme Efforts Across the Aegis Lifecycle (Core Aegis FYs 2005–2010)

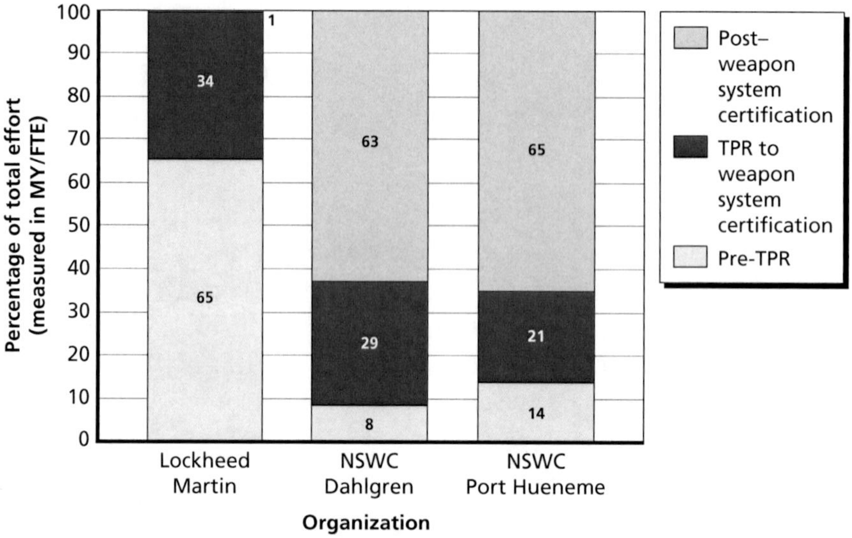

NOTE: MY = man-year; FTE = full-time equivalent. Some data are reported in MY and some in FTE.
RAND *RR161-3.1*

PEO Integrated Warfare Systems manages the combat systems for the entire U.S. Navy. The remaining systems are under the control of Space and Naval Warfare Systems Command (communication systems) and Naval Air Systems Command (Identification Friend or Foe and Tomahawk systems). PEO Integrated Warfare Systems "owns" the AWS and integrates the ACS. Aegis Integrated Combat Systems, Major Program Manager (PEO Integrated Warfare Systems 1.0), the organization that deals directly with Aegis, supports each of the lifecycle phases; it manages the weapon system and is a major contributor to the IWS business model and the Aegis open architecture.

Lockheed Martin Maritime Systems and Sensors (MS2), in Moorestown, New Jersey, is the current combat systems engineering agent for Aegis. As such, it participates in Aegis design, conducts design studies of new hardware and software as potential upgrades, coordinates and manages AWS computer program development (including the development done by subcontractors), and works to integrate and test the system. In addition, Lockheed Martin MS2 designs, engineers, and builds the Aegis SPY radar. Figure 3.1 confirms that the vast majority of Lockheed Martin Navy-funded core Aegis activity between FYs 2005 and 2010 occurred prior to baseline weapon system certification.

Aegis TECHREP, colocated with Lockheed Martin MS2 in Moorestown, works closely with Lockheed Martin during the development, integration, and testing of new Aegis software and hardware to provide government validation of software, documentation, government-furnished equipment, government-furnished information, and government-furnished computer programs. Aegis TECHREP is the on-site government representative, operates the CSEDS, and is responsible for providing technical support and leadership for Aegis. (CSEDS is discussed in greater detail in the section on facility roles and responsibilities in this chapter.) Aegis testing by Lockheed Martin MS2 and Aegis TECHREP focuses on validating contracted work. That is, testing to validate new functions in the software that Lockheed Martin MS2 is contracted to perform. Lockheed Martin MS2 and Aegis TECHREP also perform regression testing of previously developed functionality, but on a smaller scale.

Three NSWCs are involved with Aegis work: NSWC Dahlgren in Virginia and NSWC Port Hueneme and NSWC Corona in California. NSWC Dahlgren focuses on software, with some effort in hardware selection and prototyping. NSWC Port Hueneme focuses on hardware in-service support. NSWC Corona focuses on independent performance analysis.

NSWC Dahlgren is the lifetime support engineering agent for AWS and, as such, is involved with engineering; testing; combat system integration; AWS and ACS certification; program builds and fleet deliveries; and fleet support, including casualty reporting responses. As a part of the U.S. Navy Review Team, NSWC Dahlgren reviews requirements and designs from an operational perspective to ensure that operational requirements for Aegis are met. It performs government testing to assess progress during development from an operational perspective and also performs system I&T, which leads to AWS and ACS certification, using the land-based test sites at SCSC, IWSL, and CSEDS. Dahlgren, Virginia, is also the location of SCSC, which oversees training and, specifically, the Aegis Training and Readiness Center. NSWC Dahlgren manages the IWSL, a facility that provides testing, development, and fleet support (e.g., computer program builds, deliveries, and problem reconstruction of fleet issues). Figure 3.1 shows that, between FYs 2005 and 2010, Navy-funded core Aegis activity at NSWC Dalhgren was distributed across phases of the weapon system lifecycle.

NSWC Port Hueneme is the in-service engineering agent for Aegis. It supports fleet readiness testing, selected restricted availability shipyard installations, modernization, Combat System Ship Qualification Trials, and casualty reporting analysis and recovery; it also grooms hardware to ensure battle readiness. NSWC Port Hueneme is a key member of the U.S. Navy's review team during the requirements and specifications phase of system development. One of its main roles is testing new weapon hardware. It performs East and West Coast Combat System Ship Qualification Trials. NSWC Port Hueneme has a limited suite of Aegis equipment. Its primary testing roles include shipboard testing and providing readiness and maintainability expertise for developmental and certification testing. Figure 3.1 shows that, between FYs 2005 and 2010, Navy-funded core Aegis activity at NSWC Port Hueneme was distributed across phases of the weapon system lifecycle.

NSWC Corona, located in Corona, California, serves as an analytical organization for Aegis live-fire test events. Its role is to provide an independent evaluation of the warfighting effectiveness of Aegis and to recommend changes.

Facilities

Facilities play a crucial role in the development, fielding, and support of the ACS. The Aegis facilities support air defense (core Aegis) development, as well as BMD and international activity. As discussed earlier, individual facilities tend to focus on specific aspects of the development process, but all are necessary to support the Aegis enterprise. Table 3.2 summarizes the role of each facility across the Aegis lifecycle, as drawn from our interviews. Figure 3.2 shows the distribution of effort the facilities expend across phases of the Aegis lifecycle.

Table 3.2
Roles of Facilities Across Aegis Lifecycle as Surmised from Interviews

	Phase in Weapon System Lifecycle			
Facility	**Development**	**I&T**	**In-Service Support**	**Training**
NSCC	√	√		
CSEDS		√		√
SCSC		√	√	√
IWSL	√		√	

NSCC is the computing facility at Lockheed Martin's complex in Moorestown, New Jersey. NSCC has a primary role in the developmental testing phase and a secondary role in I&T. It supports Aegis, BMD, and Aegis foreign military sales. Figure 3.2 shows that the majority of Navy-funded core Aegis activity at NSCC occurs before weapon system certification, and more than a third of that activity occurs before the TPR.

CSEDS is a Navy-run I&T site located near Lockheed Martin's complex in Moorestown, New Jersey. CSEDS has a primary role in integrating and testing both U.S. and international Aegis baselines. It plays an important part in the Aegis lifecycle because it is where the Aegis computer program and hardware are integrated for the first time by a U.S. Navy organization. Since it is a developmental test site, testing can be performed sooner than if the equipment were taken to an operational test site, where it would be exposed to real radar loads and would interact with active equipment. Figure 3.2 shows that the majority of Navy-funded core Aegis activity at CSEDS occurs during the I&T effort after TPR and before weapon system certification.

SCSC is a Navy-run I&T facility in Wallops Island, Virginia, that provides a high-fidelity engineering environment for fleet support. SCSC primarily supports I&T and in-service support, with important but relatively lower loads from Aegis team training. Wallops Island is located on the coast to support unobstructed open-ocean testing. SCSC conducts live performance assessments. Combat System Ship

Figure 3.2
Distribution NSCC, CSEDS, SCSC, and IWSL Efforts Across Aegis Lifecycle (Core Aegis FYs 2005–2009)

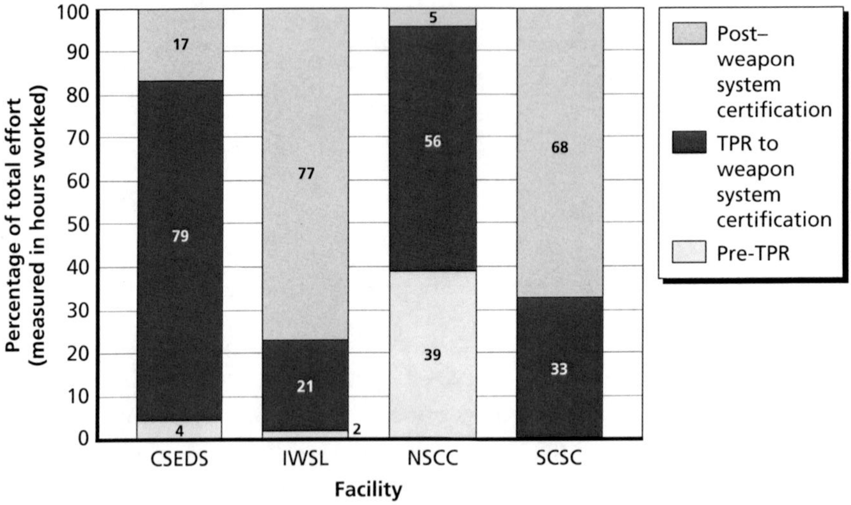

RAND *RR161-3.2*

Qualification Trials for East Coast ships are conducted in a designated operating area, which allows SCSC to provide radar coverage and support in conjunction with NSWC Port Hueneme. The permanent SCSC staff operates and maintains the equipment. The user community, in this case NSWC Dahlgren or Port Hueneme, conducts the development work. Unlike NSCC and CSEDS, SCSC does not support any international activity. SCSC is funded to operate five days a week at two shifts per day. Figure 3.2 shows that essentially all of the Navy-funded core Aegis activity at SCSC occurs after I&T has begun, and most of the activity occurs after weapon system certification.

IWSL is a Dahlgren facility that supports computer program engineering and support. It supports the lifetime support engineering agent in generating, maintaining, updating, and certifying ACS computer programs. Figure 3.2 shows that essentially all of the Navy-funded core Aegis activity at IWSL occurs after I&T and, like at SCSC, most activity occurs after weapon system certification.

Balance of Effort

Government Versus Industry

The AWS is developed through collaboration between government and industry. Our interviews suggest that, historically, government and industry work in the development process was more segregated than it is today. U.S. Navy officials reported that, in the past, the government would receive a software product and proceed to test and potentially modify the code to suit its needs (sometimes duplicating effort expended by industry developers). Today, representatives from both government and industry sit on the various integrated product teams that manage the Aegis development process. Whatever the history, the IWS business model anticipates very aggressive I&T timelines, and as we discuss later, the relationship and balance of effort between government and industry may be a source of risk.

Historically, the government level of effort has been significant. Figure 3.3 shows the distribution of effort across Lockheed Martin, NSWC Dahlgren, and NSWC Port Hueneme. During FYs 2005–2010, the NSWC exhibited more than 60 percent of the overall effort devoted to the core Aegis program (i.e., not including activity related to Aegis BMD or international activity).[4] Over the same period, the Warfare Centers accounted for 30 percent of the effort expended on activities associated with pre-TPR baselines (i.e., before the main I&T effort). For baselines between TPR and weapon system certification (the interval corresponding roughly to the most strenuous period of I&T), NSWC Dahlgren has historically expended essentially as much effort as Lockheed Martin; Dahlgren and Port Hueneme together accounted for more the 60 percent of the total effort during this period. There was some variation in government and industry levels of effort across baselines, but government organizations accounted for at least 48 percent of the overall effort expended between TPR and operational certification of BL 7.1R and ACB-08/BL 8.

---

[4] These figures underestimate the overall government role in so-called core Aegis, since the data do not include efforts by Aegis TECHREP and PEO IWS or NSWC Port Hueneme efforts funded by PEO Ships.

Figure 3.3
Distribution of Development Efforts Across Lockheed Martin, NSWC Dahlgren, and NSWC Port Hueneme (Core Aegis FYs 2005–2010)

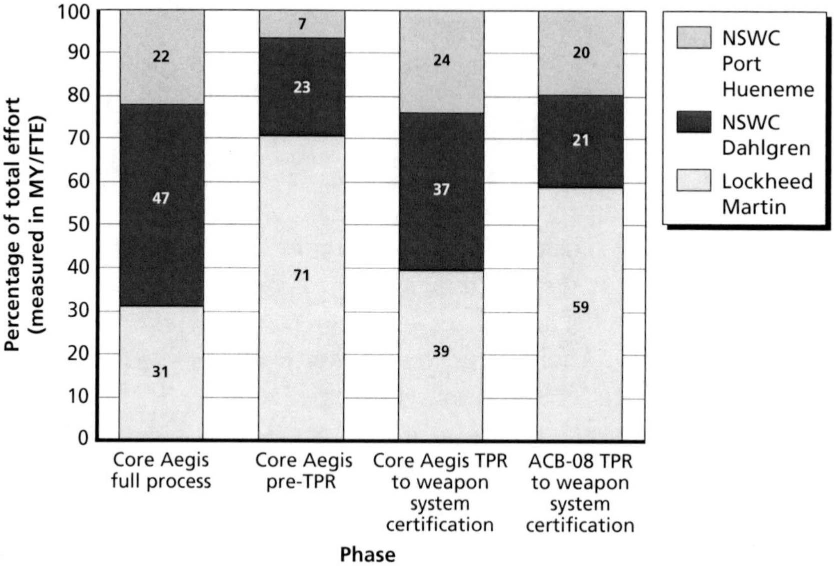

NOTE: Some data are reported in MY, some in FTE.
RAND *RR161-3.3*

Overall, these data lead us to conclude that the government elements of the Aegis enterprise have a sizable role, as measured both in relative and absolute level of effort. As we will discuss in Chapter Seven, this seems to stand in contrast to the ARCI and SSDS approaches and may be a source of risk in executing the IWS business model.

U.S. Navy, BMD, and International Activity

The Aegis enterprise has three resource sponsors: the U.S. Navy, the MDA, and foreign governments that purchase U.S. Navy weapon systems through foreign military sales. In fact, BMD and FMS appear to be driving an increasing percentage of work at Lockheed Martin and activity at key integration and test facilities. Figure 3.4 shows man-years devoted by Lockheed Martin to major development activity over time; Figure 3.5 shows the number of facility hours at CSEDS.

Figure 3.4
Trends in Lockheed Martin Level of Effort by Baseline (FYs 2001–2010)

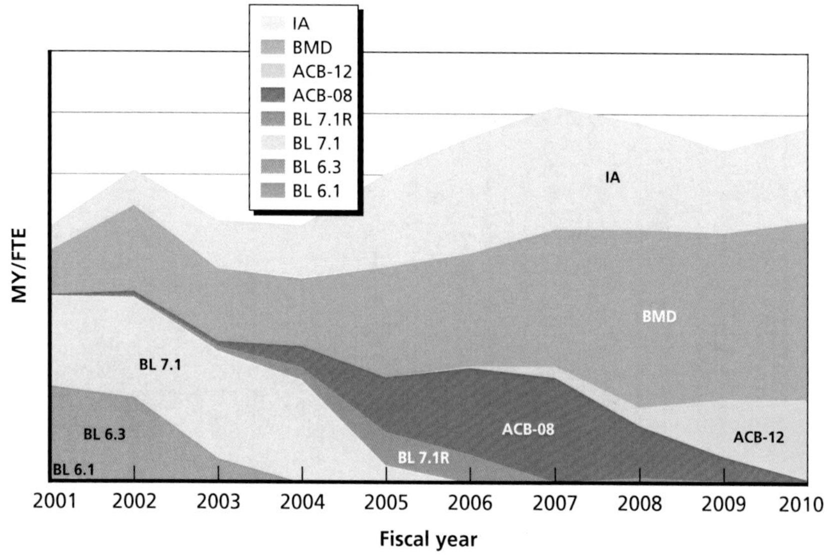

NOTE: The scale of the dependent axis is not specified out of respect for the confidentiality of proprietary data.
RAND *RR161-3.4*

All trends show an increasing industry effort expended on BMD and international activity in both absolute terms and relative to the effort devoted to core Aegis development.[5]

These trends foreshadow competition between the U.S. Navy and BMD for resources such as facility time, engineering talent, control and access to a common source library, and specification of weapon system architecture and interface definitions. The competition may pose a risk for the IWS business model, since, in an effort to meet all demands, individual users may have to make compromises on what capability gets to the fleet. We discuss this risk in a later chapter.

---

[5] The U.S. Navy remains the dominant user of IWSL and SCSC, as those facilities are not used for international activity. Our data do not allow us to assess the levels of effort the NSWCs devote to BMD and international activity.

Figure 3.5
Trends in CSEDS Usage by Baseline (FYs 2001–2009)

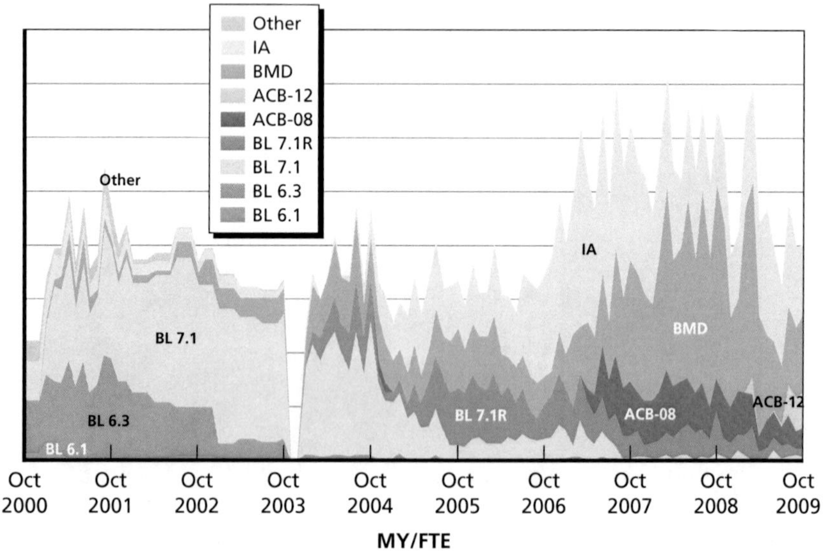

NOTE: The scale of the dependent axis is not specified out of respect for the confidentiality of proprietary data.
RAND *RR161-3.5*

Recent Versus Legacy Baselines

The IWS business model will modernize the fleet by replacing legacy baselines. The model is also expected to affect the number of in-service baselines (as will be quantified in Chapter Four). These changes may affect what resources the U.S. Navy devotes to older versus more recently developed capabilities.

In fact, facility usage due to legacy baselines is small at IWSL and SCSC and insignificant at CSEDS and NSCC. Figure 3.6 shows the distribution of effort across Aegis baselines (Core Aegis FY 2009). We see that, in recent years, BL 7.1 is the oldest baseline for which there is measurable usage at CSEDS, and BL 7.1R is the oldest in use at NSCC. IWSL and SCSC experience some demands from BLs 6.3, 6.1, 5.3 3A, and 2.10, which are generally considered legacy baselines.

Figure 3.6
Distribution of Effort Across Aegis Baselines (Core Aegis FY 2009)

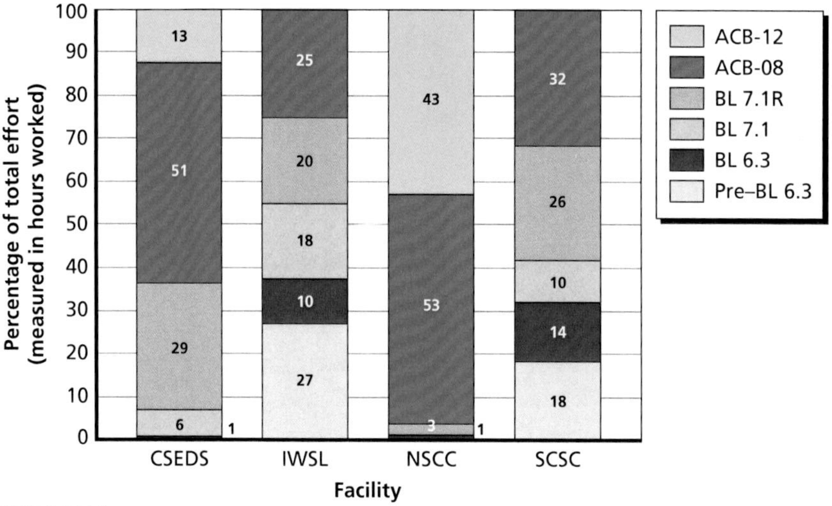

RAND *RR161-3.6*

From a workforce perspective, Lockheed Martin focuses on the recent baselines, consistent with its role as prime integrator. The historical division of NSWC manpower across facilities is more difficult to assess because government support of legacy baselines is often included in a general support category, for which man-years are not broken out by baseline. Figure 3.6 shows that these organizations do expend a considerable amount of their overall effort on in-service support. Interviews indicate that NSWC Dahlgren and Port Hueneme are the organizations that would support the activity at IWSL and SCSC—this suggests that their support activity extends, in fact, to legacy BLs 6.3, 6.1, 5.3 3A, and 2.10.

In short, the data indicate that legacy baselines are the source of a small amount of facility usage but probably a significant amount of effort by the NSWCs.

Development Versus I&T

The IWS business model decouples the development and I&T timelines, envisioning a regimented and periodic I&T effort that harvests robust development efforts running in parallel. As we will discuss later, this change may affect the balance of resources devoted to development versus I&T, as well as the organizations and facilities that contribute to those efforts.

Figures 3.1 and 3.2 show how we estimate the distribution of core Aegis enterprise effort (FYs 2005–2010) and facility usage (FYs 2001–2009) among the main efforts of development, I&T, and in-service support. We see that periods of development account for the lowest percentage of facility usage overall. NSCC and CSEDS are more active during I&T, whereas IWSL and SCSC are more active post-certification. Lockheed Martin's efforts are predominantly dedicated to development and I&T. NSWC Dahlgren exhibits effort across categories, and NSWC Port Hueneme concentrates on in-service support, along with some I&T activites. From data not shown here, we can see that Lockheed Martin's efforts appear to fluctuate between development and I&T over time, mostly in accordance with the progression of development milestones. For NSWCs Dahlgren and Port Hueneme, the data show a relatively consistent balance of effort across lifecycles over time.

Summary

In this chapter, we have summarized the self-described roles and responsibilities of organizations and test facilities in what is a vast Aegis enterprise, as well as analyzed historical manpower and facility usage. From this, we can draw several conclusions. First, the government plays a very large role in the Aegis enterprise—with five separate government organizations involved at a significant level of effort in every phase of the weapon system lifecycle. Second, BMD and international activity represent an increasing demand on the Aegis enterprise, suggesting that, in the future, the U.S. Navy may have diminishing leverage on its people, processes, and facilities. Finally, there appears to be a relatively small, but not insignificant, level of effort devoted to legacy baselines.

CHAPTER FOUR

Impact of the IWS Business Model and Implementation Choices on the Fleet

PEO Integrated Warfare Systems has a range of policy choices to make that will affect both the technical infrastructure required for Aegis development and the capabilities delivered to the Aegis fleet. It is necessary to articulate the effect of these choices on the fleet and on development requirements. The PEO can choose the pace of ACB and TI upgrades and can choose to have individual ships receive either every upgrade or every other upgrade. Also, the PEO can choose whether to field ACB and TI upgrades simultaneously or to stagger them.

In this chapter, we quantify the impact of the IWS business model on the Aegis fleet. To this end, we developed a model that tracks individual ships over time as they are modernized. This approach enabled us to aggregate and measure the effect of different parameters of the IWS business model on individual ships and fleet-wide capabilities.

The RAND Dynamic ACB/TI Model

To gain a better understanding of the impacts of different ACB/TI schedules on the Aegis fleet, we developed a predictive model to forecast the future baseline or upgrade composition of the fleet.[1] For a given upgrade schedule, the model tracks each ship's (current and future) configuration throughout its entire lifecycle under a variety of assumptions. Currently, the model is written to track the fleet transi-

[1] The model is a discrete event and time-step program written in the Mathematica software programming language.

tion from the legacy Aegis baseline framework to the ACB/TI configuration. However, it can easily be expanded to track other changes (Aegis-related and not), such as minor upgrades and modifications or equipment-specific changes, as they spread throughout the fleet.

Model Inputs and Assumptions

Currently, the key inputs that define each ACB/TI upgrade schedule are as follows:[2]

- *ACB drumbeat:* the number of years between successive ACB developments that enter the fleet
- *ACB lag:* the number of ACB developments between individual ship upgrades (e.g., an ACB lag of 1 implies that each ship receives *every* ACB; an ACB lag of 2 implies that each ship receives *every other* ACB)
- *TI drumbeat:* the number of years between successive TIs that enter the fleet
- *TI lag:* the number of TI developments between individual ship TI upgrades
- *ACB/TI offset:* the number of years that offset the ACB and TI development and fielding schedules (after ACB/TI-12).

Figure 4.1 compares the fielding schedules for three ACB/TI modernization options defined by the following parameters:

1. a new ACB every four years and a TI every eight (2/2, 4/2, 0)
2. a new ACB and TI every four years (4/1, 4/1, 0)
3. a new ACB and TI every four years, but offset by two years (4/1, 4/1, 2).

Unless otherwise specified, the model assumes that each ship receives its scheduled upgrade on time. It does not predict future ship availability for such upgrades. The model does allow for inputs that

---

2 The IWS Combat System Acquisition Plan calls for a four-year ACB drumbeat and a four-year TI drumbeat, with individual ships receiving every other upgrade. The plan does not have an offset (i.e., ACB and TI upgrades are developed and fielded simultaneously).

Figure 4.1
Graphical Comparison of Fielding Schedules for Three Potential ACB/TI Upgrade Options

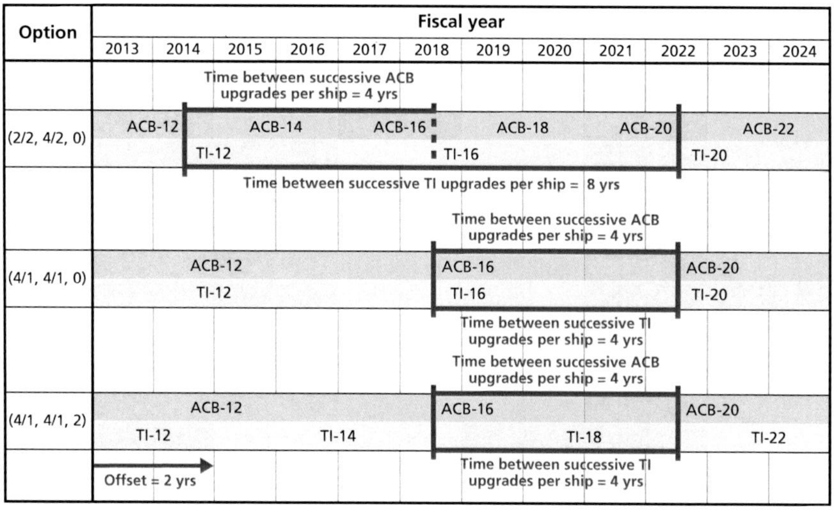

RAND *RR161-4.1*

restrict the total number of upgrades allowed per year, which could delay some ship upgrades. Most of our assumptions concerning the future Aegis fleet are from *U.S. Navy Force Structure and Shipbuilding Plans: Background and Issues for Congress* (O'Rourke, 2012) and the *Report to Congress on Annual Long-Range Plan for Construction of Naval Vessels for FY 2013* (U.S. Navy, 2012). After FY 2035, we assume that *two* Aegis ships will be commissioned per fiscal year. The model can project the Aegis force structure well beyond FY 2035, assuming the Aegis program remains the program of record.

According to an April 2012 report to Congress (U.S. Navy, 2012), all Flight IIA DDG 51s (DDGs 79–121) will have an extended service life of 40 years (up from 35 years). We assume that the Flight III DDG 51s (DDG 122 and higher) will also have a 40-year service life.[3] The remaining Aegis ships—CG 47s (CGs 52–73), Flight I DDG 51s (DDGs 51–71), and Flight II DDG 51s (DDGs 72–78)—are currently

---

[3] Flight III DDG 51s are the third variant of the *Arleigh Burke* destroyer to enter service.

scheduled to retire after 35 years of service. Several ships have already received or are scheduled to receive their major modernization, taking them from a legacy baseline configuration to the ACB/TI construct. The first seven ships (CGs 52–58) were to receive ACB/TI-08. Under the IWS business model, these ships will receive no more significant upgrades for the remainder of their service life.

According to representatives from PEO IWS, one DDG (DDG 79) will begin its modernization upgrade in FY 2016, and from FY 2020, three DDGs will begin the modernization process each fiscal year. At that rate, the last legacy baseline ship, DDG 112, will become ACB/TI-configured in about FY 2028. Consequently, there will be a total of approximately 50 cruiser and destroyer modernizations from FY 2009 to FY 2028. All ships that follow DDG 112 will enter the fleet in the ACB/TI configuration. We assume that these ships will receive the current ACB and TI *1.5 years* before commissioning and that they receive their first upgrade based on their respective initial install dates. For example, if the schedule calls for an upgrade once every four years, DDG 113 (and higher) will receive its first upgrade 2.5 years after its commissioning and additional upgrades every four years thereafter. Finally, we follow the five-year rule stating that no ship shall receive an upgrade within five years of its retirement date.[4]

Model Outputs

In this section, we present a series of model outputs using the current IWS ACB/TI upgrade schedule of a four-year ACB drumbeat and TI drumbeat, with individual ships receiving every other upgrade. Figure 4.2 shows the number of ships in the fleet by baseline, ACB, and TI over a period of 50 years. Figure 4.3 displays the number of upgrades, and Figure 4.4 plots the number of ACB/TI configurations in the fleet over the same period. Figure 4.5 quantifies the software and hardware age distributions while the fleet transitions from the legacy baseline to the AC/TI framework. Software and hardware age are dis-

[4] Note that most of the values and preplanned schedules discussed in this section are parameters in the model and, as such, can be modified as needed.

cussed in more detail later in this chapter. Finally, Figure 4.6 shows the expected delay between software development and fleet installation.

Figure 4.2 plots the number of ships in the fleet by legacy baseline and ACB (left side) and by legacy baseline and TI (right side) for the IWS model for FYs 2010–2060. The total Aegis force peaks at 90 ships in 2020, then begins to decline steadily as the Flight I and II DDGs retire from service, and bottoms out at 77 ships around FY 2033. The force then begins to expand again due to the 40-year service time of the later DDGs. The oscillation that occurs between 2040 and 2060 mirrors a similar oscillation in procurement 40 years earlier. Based on our procurement assumptions, the Aegis force will continue to increase past 2060, until it reaches about 90 ships.

Several observations about the Aegis fleet are relevant as it transitions to the IWS business model. First, the legacy baselines remain in the fleet for a long time; the last legacy baseline ship does not leave the fleet until 2035. Until then, the UYK-43 computing hardware must be maintained. The effects of the legacy model on the fleet are independent of any IWS choices with regard to ACB/TI upgrade periodicity. The cruiser and destroyer Aegis modernization (AMOD) rate determines when the legacy baselines leave the fleet. Second, the fleet composition in any given year can be determined by reading the vertical axis. In both the software (ACB) and hardware (TI) cases, the number and age of the deployed systems are improved relative to the legacy business model.

Some examples from Figure 4.2 of the future fleet composition under the IWS plan include the following:

- FY 2020: There are 86 Aegis ships in the fleet, 54 of which remain under the legacy baseline system; the remaining 32 ships have already entered the ACB/TI construct. Specifically, there are
  - 22 BL 7.2, 5 BL 6.3, 1 BL 6.1, 25 BL 5.3, and 1 BL 3.1 ship
  - 7 ACB-08/TI-08, 14 ACB-12/TI-12, 5 ACB-16/TI-16, and 6 ACB-16/TI-16 ships.
- FY 2050: There are 83 Aegis ships in the fleet, and all are ACB/TI ships. Of these, there are 1 ACB-36/TI-32, 7 ACB-40/TI-36,

Figure 4.2
Aegis Ship Count by ACB and TI Version for the IWS Modernization Plan (FYs 2010–2060)

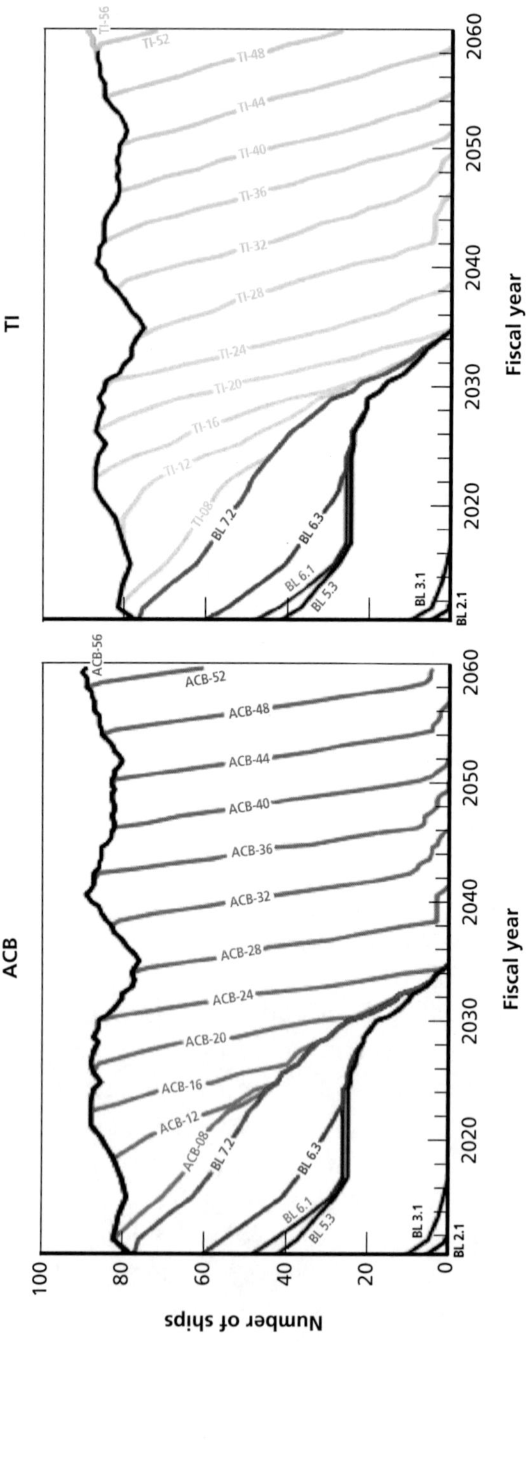

RAND *RR161-4.2*

3 ACB-40/TI-40, 33 ACB-44/TI-40, and 39 ACB-44/TI-44 ships.

Figure 4.3 plots the number of Aegis modernizations and the number of ACB and TI upgrades for ships already ACB/TI-configured. Some ACB ships begin receiving their first ACB upgrades in 2016—specifically, legacy baseline ships that were originally modified to the ACB-12 standard.[5] The total number of ACB-to-ACB upgrades from 2010 to 2060 is 242, averaging 4.8 per year, whereas the average during the final 10 years of the period (FYs 2050–2060) is 9.9 upgrades per year. Upgrade frequency is consistent at ten per year once the fleet reaches the 90-ship level.

Figure 4.3
Number of Major Modernizations, ACB Upgrades, and TI Upgrades per Year Under the IWS Model (FYs 2010–2060)

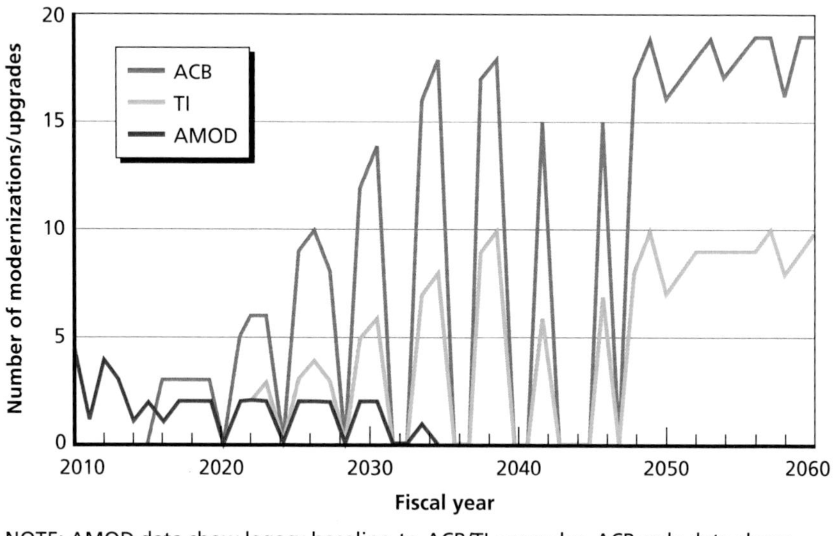

NOTE: AMOD data show legacy baseline-to-ACB/TI upgrades, ACB only data show ACB-to-ACB upgrades, and ACB and TI data show complete (ACB + TI) upgrades.
RAND *RR161-4.3*

[5] Our model assumes that the Navy will provide the resources for all modernizations and upgrades. The modeling results are meant to assess policy decisions available to the Navy, not the stability of funding.

Figure 4.4
Number of Active Software, Hardware, and Software-Hardware Configurations in the Fleet Under the IWS Model (FYs 2010–2060)

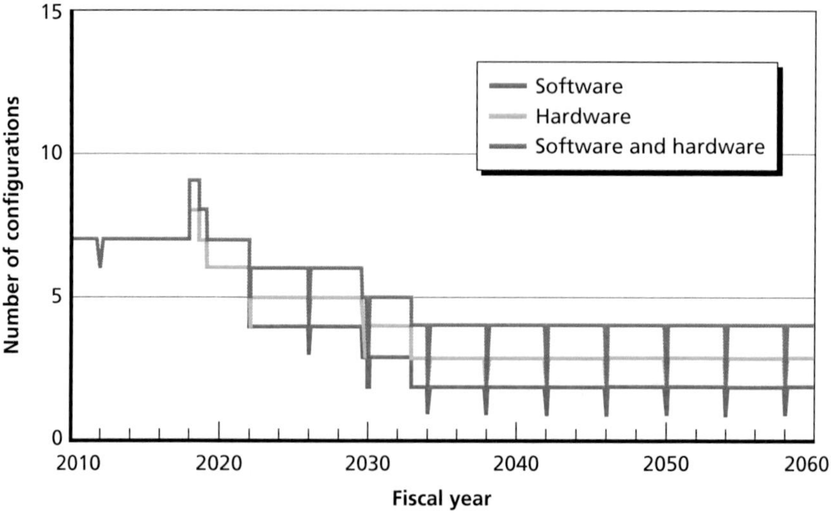

NOTE: The figure does not include additional configurations for ships within five years of retirement.
RAND *RR161-4.4*

Figure 4.4 shows the number of active configurations—software (baselines and ACBs), hardware (baselines and TIs), and software-hardware combinations (baselines and ACB/TIs)—in the fleet under the IWS model at any given time until FY 2060. The number of active configurations is important because it will drive much of the demand for training and in-service support. For this reason, Figure 4.4 excludes ships that are within five years of retirement, as their training and in-service support requirements diminish significantly. The duration and complexity of the transition from the legacy to the IWS business model is evidenced by the fact that the number of configurations actually increases initially and does not reach a steady state until 2033. During the transition period, the number of software-hardware combinations reaches a maximum of eight as new ACB/TI configurations enter the fleet while legacy ones die out. During the steady-state period (FYs 2050–2060), however, the number of configurations drops and appears to level out at an average value of 2.9. The number of configurations in the steady state is determined by the choice of ACB and TI drumbeats and individual ship upgrade decisions.

Finally, we consider how the IWS business model affects software and hardware age distributions in the fleet. We measured how quickly a particular software development propagates throughout the fleet. For each ACB and TI, we assume that development is complete by the beginning of the year for which it is named. For example, development for ACB-12 and TI-12 should have been complete by October 1, 2011, and this is the date from which their age is measured. The underlying assumption is that it takes a year to integrate and test new software and hardware. In the case of ACB/TI-12, the upgrade begins entering the fleet at the start of FY 2013. Meanwhile, FY 2012 development is geared toward the next ACB and TI iterations.

We estimated the legacy baseline ages based on when they entered the fleet and assumed a similar I&T time.[6] Figure 4.5 plots both the maximum and average software and hardware ages in the Aegis fleet. Once again, we exclude ships that are within five years of retirement and that are no longer eligible to receive upgrades. The maximum software and hardware ages drop dramatically around FY 2028, when the last legacy baseline ship is upgraded to the ACB/TI framework. From FY 2050 to FY 2060, the median values for the maximum and average ages are 12.6 and 8.0 years, respectively.

Impact of the IWS Business Model on the Aegis Fleet

We conducted our assessment of the IWS business model's impact on the Aegis fleet in two parts. First, we compared the characteristics of the Aegis fleet under the legacy and IWS business models. Then, assuming the incorporation of the IWS business model, we evaluated the pertinent policy choices that will be made by PEO Integrated

---

6 The legacy baseline dates assumed here are BL 2.1: March 1985, BL 3.1: November 1987, BL 5.3: August 1995, BL 6.1: November 1998, BL 6.3: July 2000, BL 7.1: June 2002, and BL 7P1R: May 2007.

Figure 4.5
Maximum and Average Software and Hardware Ages in the Active Fleet Under the IWS Model (FYs 2010–2060)

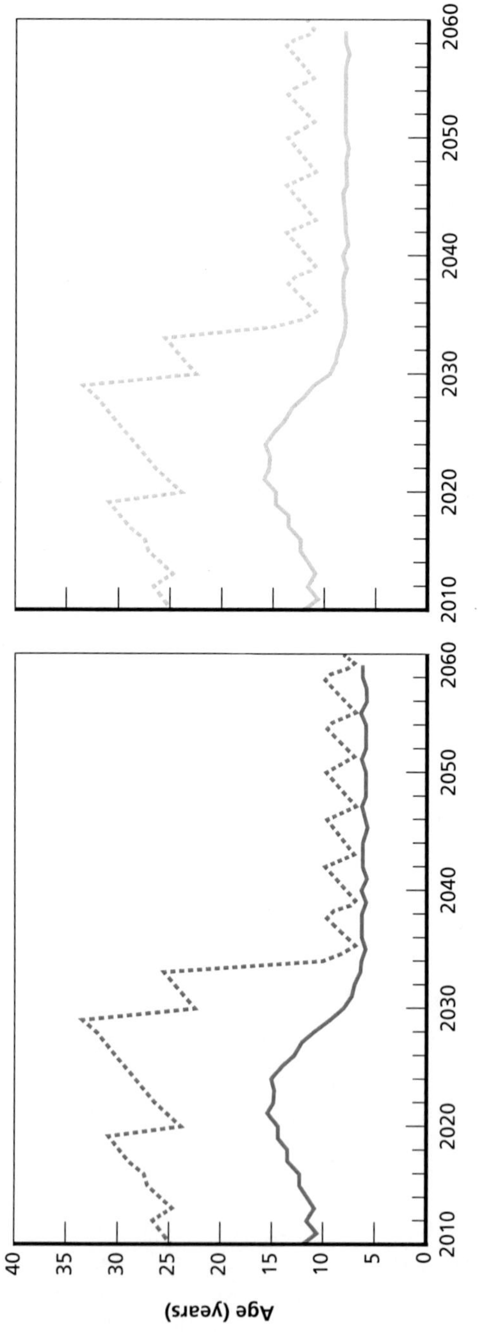

RAND *RR161-4.5*

Warfare Systems. In evaluating the legacy and IWS business models, we assumed that the development cycle would be the same for both. Under the legacy business model, only those ships entering the fleet by means of new construction or those undergoing their midlife overhaul receive the upgrade. This comparison isolates the effect of the OA business model from potential improvements due to an increased development pace.

ACB and TI drumbeats are development choices made to provide capabilities to the fleet. In this case, the proxy for capability is the age of either the computer hardware or software. Implicit in this assumption is that newer technology, whether software or hardware, performs better than older technology. In addition to measuring capability, the model provides insights into the mixture of deployed Aegis upgrades and the modernization churn in the fleet. The fleet mixture metric is the number of ACB, TI, and ACB/TI combinations in the fleet. The number of TI and ACB upgrades required per year is the metric for modernization churn. What this model does not illuminate is how these choices affect the technical infrastructure required to support such development. It does, however, provide some inputs that can be used to examine these effects, such as the pace of development. The impact on technical infrastructure is addressed later in this report.

Legacy Versus IWS Business Model

The Aegis program has thrived on evolutionary development since its inception. RADM (ret.) Wayne E. Meyer focused the program around sound system engineering practices and the motto "Build a little, test a little, learn a lot." As a result, the Aegis program has fielded a new baseline every five to six years since its introduction. From a development standpoint, the differences between the legacy and IWS business models are minimal. However, the IWS model allows introduction of software and hardware improvements on any modernized Aegis ship, whereas the legacy system only made improvements on the version of software being developed. We assume that the legacy business model is consistent with the way the Aegis program has historically operated, with the exception of the incorporated midlife overhaul. For the purposes of this analysis, we assume a new baseline (legacy model) or a new ACB (IWS model) every six years. As pointed out earlier, under the legacy model, a ship receives an initial Aegis and an updated version at the midlife upgrade. Under the IWS business model, a new ACB is available every six years and is installed on every ship.

Implementation under the IWS model dramatically alters the performance characteristics of the Aegis fleet. Figure 4.6 shows the fleet makeup from FYs 2010 to 2060 under the IWS business model (left side) and under the legacy business model (right side). The aggregate Aegis fleet statistics for the legacy and IWS cases are shown in the first and second row of Table 4.1, respectively. The average age of the hardware and software under the legacy business model is 14 years, and the maximum age for both is 25.6 years. Under the IWS business model, the average age drops to 8.1 years and the maximum age to 11.8 years.

The investment made in transitioning the fleet dramatically reduces the age of the deployed technology. Under the IWS business model, individual ACBs and TIs stay in the fleet for shorter periods. As a result, there are fewer versions of Aegis in the fleet that need to be supported. In the legacy case, an average of 4.3 Aegis versions are deployed at any given time. Under the IWS model, that number falls to two. Thus, the IWS business model reduces both the age of the technology and the number of Aegis versions deployed simultaneously.

The implementation of OA in Aegis also leads to cost efficiencies. Under the legacy business model, each upgrade is installed on about 21 percent of the Aegis fleet. In the IWS case, each upgrade is installed on about 83 percent of the fleet.[7] Essentially, the development investment is spread across four times as many ships. When measured at the fleet level, the power of the IWS business model to improve deployed capabilities is impressive.

However, there is a measurable cost to providing new software and computer hardware capabilities that spread rapidly across the Aegis fleet. Under the legacy business model, ships receive their combat system during new construction and the midlife overhaul. Individual ships receive no additional upgrades and, therefore, incur no additional costs. Under the IWS business model, ships are constantly being

[7] Ships within five years of retirement are not upgraded.

Figure 4.6
Aegis Fleet Composition Under the IWS and Legacy Models (FYs 2010–2060)

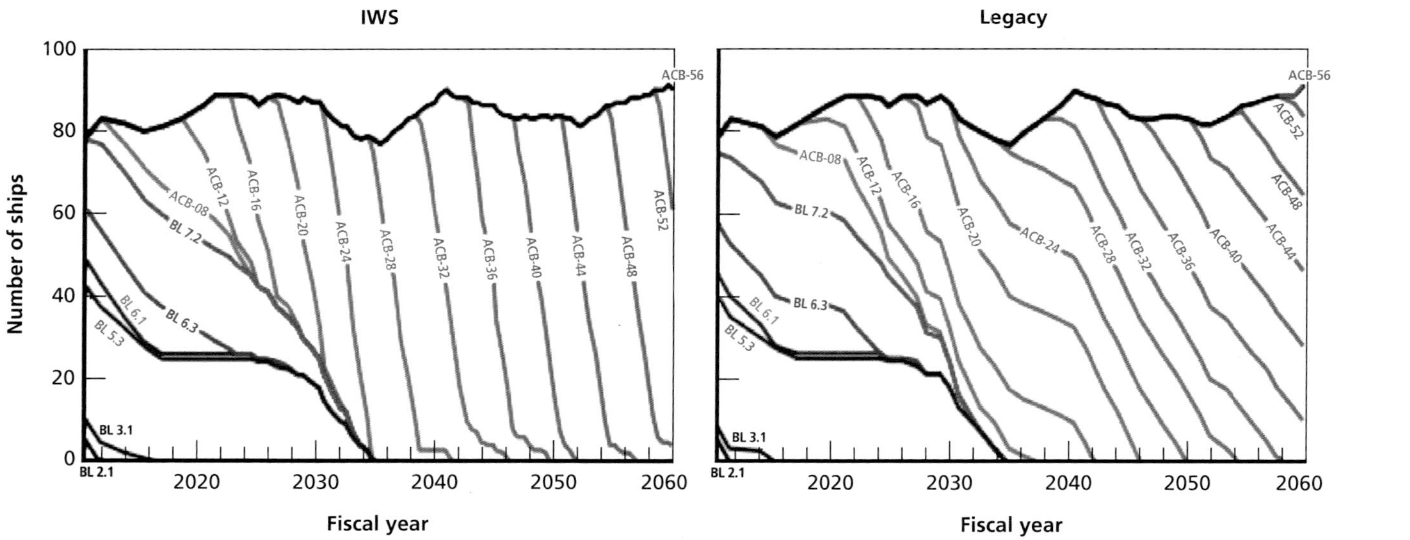

RAND *RR161-4.6*

Table 4.1
Dynamic ACB/TI Model Results

Schedule	ACB Rate	ACB Lag	TI Rate	TI Lag	Percentage of Ships per Baseline	Upgrades per Year		Number in Fleet			Software Age		Hardware Age	
						ACB	TI	ACB	TI	ACB/TI	Max.	Avg.	Max.	Avg.
Pre-OA					21			4.3	4.3	4.3	25	14.2	25	14.2
Option 1	6	1	6	1	84	11.3	11.3	2	2	2	11.8	8.1	11.8	8.1
Option 2	4	2	4	2	44	9.3	9.3	3	3	3	12.6	8	12.6	8
Option 3	4	1	4	1	83	18.7	18.7	2	2	2	8.7	6.2	8.7	6.2
Option 4	2	1	4	1	86	39	18.7	2	2	2.9	5.5	4.2	8.7	6.2
Option 5	2	1	2	1	86	39	39	2	2	2	5.5	4.2	5.5	4.2
Option 5	2	2	2	2	41	18.7	18.7	3	3	3	7.5	5.2	7.5	5.2
Option 6	2	2	4	2	41	18.7	9.3	3	3	5.9	7.5	5.2	12.7	8
Option 7	4	1	4	1	86	18.7	18.7	2	2	3	8.7	6.2	8.3	6.2

NOTE: Shading denotes ACB/TI offset cases.

updated with the latest ACB and TI. In this case, a six-year drumbeat implies that, on average, one-sixth of the fleet will be upgraded each year. Each upgrade to an individual ship incurs cost, both for the hardware and for the team required to install the upgrade.

Changing from the legacy to the IWS business model results in a surface fleet with newer technology, a development effort that efficiently spreads its results across a larger user base, and a lower level of required support due to fewer deployed versions of the Aegis system. However, these improvements require individual ACB and TI versions to be installed on ships periodically over their lifetime. This generates a cost that is not present in the legacy business model.

Effect of ACB/TI Choices on Aegis Fleet Capabilities

The IWS business model improves overall Aegis fleet capability by spreading individual upgrades to all or portions of the fleet. The next question that needs to be answered is how PEO Integrated Warfare Systems policy choices affect those capabilities. Our dynamic model allows these choices to be evaluated at the fleet level. In this section, we evaluate the effects of ACB and TI drumbeats of two, four, and six years. Individual ships can receive either each upgrade or every other upgrade. We also address the impact of offsetting the ACB and TI upgrades.[8] Selecting the proper ACB and TI drumbeat requires PEO Integrated Warfare Systems to consider the benefits of fielding improvements to the fleet, as well as the costs and feasibility of an accelerated development schedule. In this section, we measure the benefits in terms of the average and maximum technology age in the fleet. The potential costs are captured in two metrics—the number if ship installations and the number of ACB-TI combinations. The feasibility of developing, integrating, and testing individual upgrades in an accelerated timeframe is addressed in Chapter Five.

---

[8] Offsetting ACB and TI upgrades means that the upgrades are not developed simultaneously. For example, consider the case of an offset four-year ACB and TI drumbeat schedule: An ACB upgrade is followed two years later by a TI upgrade. Both the ACB and TI upgrades are on a four-year schedule, but they are offset by two years.

The results of our dynamic ACB/TI model are shown in Table 4.1. As the drumbeat accelerates from every six years to every four years, the average technology age decreases from 8.1 years to 6.2 years, and the maximum age decreases from 11.8 years to 8.7 years. As the drumbeat further accelerates from every four years to every two years, the average technology age decreases from 6.2 years to 4.2 years, and the maximum age decreases from 8.7 years to 5.5 years. It is apparent that the largest drop in technology age occurs when the IWS business model is implemented. As the drumbeat quickens, the improvement in technology age is proportional to the increase in frequency of the upgrades. The number of upgrades that need to be fielded increases dramatically as the drumbeat quickens. As the drumbeat quickens from six years to two years, the number of ships that must be upgraded each year increases from an average of 11.3 to 39.

The impact of individual ships receiving each upgrade versus every other upgrade is significant. For example, when the ACB drumbeat is two years but a ship receives only every other upgrade, the average age of the software increases from 4.2 years to 5.2 years, and the maximum age increases from 5.5 years to 7.5 years. Meanwhile, the number of combinations deployed in the fleet increases from two to three. It is interesting to look at the difference between a four-year drumbeat with ships receiving every upgrade and a two-year drumbeat with ships receiving every other upgrade. The average age of technology falls from 6.2 years to 5.2 years and the maximum age from 8.7 years to 7.5 years. While the development infrastructure must integrate and test an ACB twice as quickly, the impact on deployed fleet capabilities is minimal. In addition, the number of combinations present in the fleet increases from two to three, potentially making in-service support more challenging.

The articulated IWS business model calls for four-year ACB and TI drumbeats, with individual ships receiving every other upgrade. As a result, individual upgrades are installed on about 44 percent of the ships. This rate is higher than under the legacy business model (21 percent), but is about half as efficient than if ships were to receive every upgrade. The average age of the software and hardware is eight years. The IWS model calls for about nine installs per year. This results in three ACB/TI combinations deployed in the Aegis fleet, on average.

Impact on the BMD Fleet

Shifting the AWS to OA and a regimented upgrade schedule, as articulated by the IWS business model, does not directly affect the size of the BMD fleet. As discussed earlier, ACB-12 and later versions combine air and missile defense functionality. Therefore, the pace at which ships are upgraded to the ACB model as part of the AMOD program, along with the upgrade rate for legacy BMD-capable Aegis ships, determines the number of BMD ships. Figure 4.7 shows the total Aegis force structure and the number of BMD ships under legacy and IWS business models. In all cases, the BMD-capable Aegis fleet rapidly expands.

The IWS business model for Aegis will provide the same benefits to BMD performance as it does to air defense. Improvements in BMD capabilities will propagate through the Aegis fleet efficiently, investments in BMD development will be spread across a larger user base,

Figure 4.7
BMD-Capable Aegis Fleet Composition Under the Legacy and IWS Models (FYs 2010–2060)

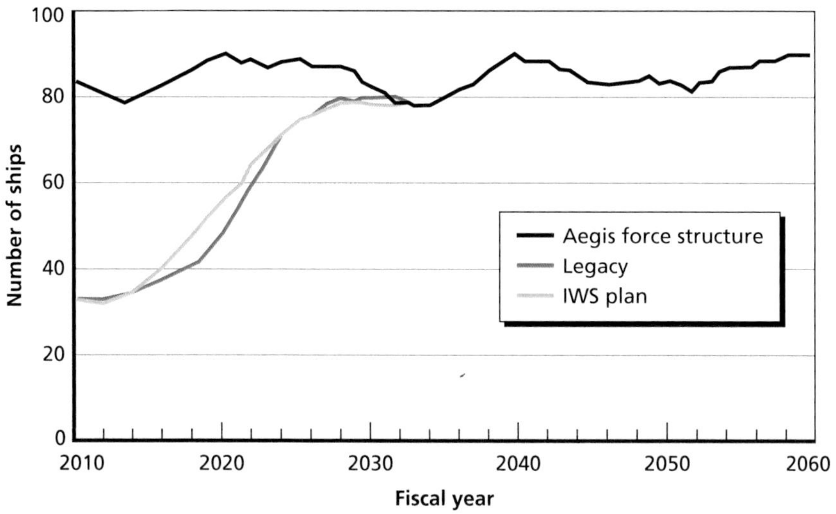

RAND *RR161-4.7*

and fewer versions of BMD will be present in the fleet. Because BMD and Aegis will operate with a single CSL, BMD performance should benefit from the investments made by the Aegis program. This will work in reverse as well, so improvements made in common functionality by BMD upgrades will enhance air defense. Finally, the transition to higher-capacity commercial computing hardware will eventually improve BMD performance.

Summary

In this chapter, we examined how the IWS business model and implementation choices affect Aegis capabilities. The transition to an OA-based approach under the IWS model has several important consequences. The capabilities and costs of a development effort are spread over many more ships. Across the Aegis fleet, the number of baselines and the age of the technology are reduced by half. Under the IWS model, introducing ACB and TI upgrades faster increases the number of upgrades per year that the fleet must support, but decreases the software age in the fleet. As the rate of modernization increases, however, it has a diminishing return on reducing technology age. If the fleet can only support a minimum number of ACB/TI upgrades per year, it has the effect of increasing technology age and the number of deployed configurations on a fleet-wide basis. Finally, the number of BMD-capable Aegis ships is independent of the ACB/TI upgrade rate; it depends only on the destroyer and cruiser AMOD rate.

CHAPTER FIVE

Implications for the Aegis Enterprise

The purpose of this chapter is to explore how the Aegis enterprise may affect, and be affected by, the IWS business model and its associated implementation choices. Our focus is on the effects of ACB and TI intervals and ACB size, since these are the choices that concern the development enterprise.[1] First, we discuss the relevance of the Navy's most recent development experiences for the purpose of understanding future demands. Then, we explore the consequences of the IWS business model for the development enterprise and discuss the trade-offs the Navy will confront as a result. Finally, we develop projections of demand on the Aegis workforce and facilities as a function of ACB size and drumbeat.

Relevance of Recent Experience in Developing the AWS

The Navy has extensive experience developing the AWS and detailed data on historical facility usage. One approach to assessing the impact of the IWS business model on the development enterprise would be to first develop a model of how historical development activity has affected the enterprise and then to extrapolate that model under anticipated future conditions. However, there are several reasons to believe that the Navy's future will be unlike its recent past; thus, it is unlikely

[1] The pace of modernization and ship upgrade frequency is relevant to the deployment of developed capability and is not expected to affect development efforts.

that such an approach would apply to the future foretold by the IWS business model.

The first decade of the 21st century featured a "clone-and-own" development process (discussed in Chapter Two), with multiple baselines in concurrent I&T. Figure 5.1 depicts the development timelines between FY 2000 and FY 2010.[2] Often during this period, two baselines were engaged in I&T and multiple baselines were concurrently in development. These concurrent baselines potentially involved similar code, since ACBs/baselines were cloned at their start from previous efforts. As Figure 2.1 showed, the IWS business model assumes serial and noncurrent I&T and fielding, as well as concurrent development with a CSL. Thus, we see that the prescribed development process differs from reality, as shown in Figure 5.1.

Figure 5.1
Aegis Development Timelines in 2000s

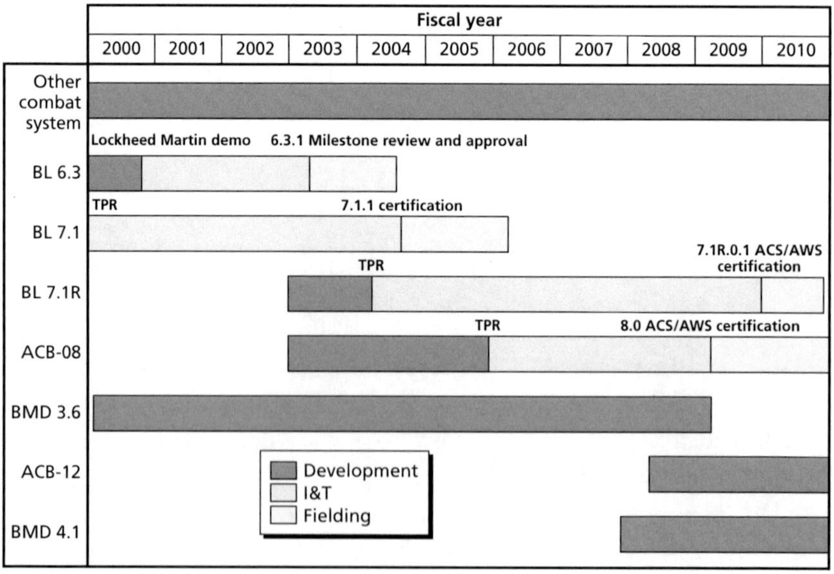

RAND *RR161-5.1*

[2] This timeline was generated using PEO Integrated Warfare Systems data on when respective baselines transitioned from development to I&T.

The substance of weapon system development activity between FY 2000 and FY 2010 also differed from the development anticipated under the IWS business model. Table 5.1 summarizes the main features of ACB-08, BL 7.1R, BL 7.1, and BL 6.3—the baselines certified between FY 2000 and FY 2010. These baselines focused on transitioning the AWS to COTS computing hardware, developing and implementing an open software architecture, and integrating combat system upgrades, including radars. We also see that previous development efforts included simultaneous hardware and software upgrades. As described in Chapter Two, the IWS business model anticipates that the fleet will complete its transition to COTS and OA in ACB-12, making way for future ACBs to focus on developing new weapon system capabilities (including BMD). The model also anticipates separating hardware and software upgrades into TIs and ACBs, leveraging a robust middleware. Thus, the substance of development in FYs 2000–2010 likely differs from future development activity.

Integration and testing involved only a relatively small development effort between FY 2000 and FY 2010, whereas the IWS business model anticipates robust development activity that will dominate I&T. Figures 3.1 and 3.2 show how enterprise effort and facility usage were

Table 5.1
Recent Aegis Baselines and ACBs

Baseline/ACB	Software Architecture	Computing Hardware	Integration of ACS Upgrades
ACB-08	Display, SPY radar, weapon control system, computer network defense OA, and OA system manager	COTS (CR 2 TI-08)	AN/SPY-1A radar; others
BL 7.1R	Display OA	COTS (CR 1)	
BL 7.1		COTS (CR 0)	AN/SPY-1D(V) radar
BL 6.3		Adjuncts added to MILSPEC	ESSM; others

SOURCE: Derived from PEO Integrated Warfare Systems data.

NOTE: ESSM = Evolved Sea Sparrow Missile.

previously dominated by baseline activity post-TPR, during the main thrust of I&T.

This recent focus on I&T might be explained by the fact that the first decade of the 21st century can be characterized as a period of transition—transition to COTS and OA. It may be that these activities were test-intensive. Whatever the explanation, the point is that the historical balance between I&T and development is unlike the future as projected by the IWS business model. Table 5.2 summarizes some of the features that distinguish the Navy's recent Aegis development experience from its future under the IWS business model.

Impact of Choices on Development

Feasible Upgrade Frequencies

The Navy's recent experience developing ACB-08 and BLs 7.1R, 7.1, and 6.3 suggests that there is some minimum amount of time necessary to integrate and test a software or hardware upgrade. Table 5.3 shows the number of calendar months that we estimate were devoted to integrating and testing the four baselines developed between FY 2000 and FY 2010. We see that 29 calendar months (roughly 2.5 calendar years) is the shortest I&T time among the four most recent baseline/ACBs. We note that BLs 6.3 and 7.1R and ACB-08 are rather similar when compared using the equivalent source lines of code (ESLOC) metric, a standard measure of effort to develop/upgrade software for

Table 5.2
Comparison of Historical Aegis Development to IWS Business Model

FYs 2000–2010	IWS Business Model
Clone-and-own development process	Concurrent development with CSL
Two ACB I&T events driven by development timelines	Single simultaneous I&T event on a regimented and periodic schedule
Focus on transiting to COTS and OA and integrating combat system upgrades	Focus on developing new weapon system capability (e.g., BMD)
Integration and testing dominates relatively small development effort	Robust development effort feeds limited I&T

Table 5.3
Duration of Main Thrust of Integration and Testing for Recent Aegis Baselines and ACBs

Baseline/ACB	I&T Time (months)	Total Facility I&T Usage (thousands of hours)
ACB-08	31	37.3
BL 7.1R	65	44.5
BL 7.1	57	89.6
BL 6.3	29	41.0

the four most recent baselines and ACBs, and that BL 7.1 stands as a unique outlier.

Recent baselines also suggest that the time allowed for I&T must align with the size and complexity of the upgrade. Figure 5.2[3] shows how facility usage between TPR and weapon system certification varies with a normalized[4] measure of ESLOC.[5] Each point corresponds to a different baseline/ACB.[6] The *total* facility hours during the main thrust of I&T across CSEDS, IWSL, NSCC, and SCSC generally increases with ESLOC. One exception is SCSC, which we will return to later. These facility data are consistent with the assumption that larger and more complex ACBs will require more effort to integrate and test.

Available manpower data are sparse and do not include BL 7.1, which is much larger than other recent efforts and therefore would be informative. However, the data presented here show that Lockheed Martin manpower is distributed as expected, with greater effort expended on baselines with more ESLOC. All things considered, we expect that larger or more complex ACBs (as measured by ESLOC) will require more effort to integrate and test.

---

3 The scale of the dependent axes is not specified because of the confidentiality of proprietary data.

4 ESLOC is normalized so that the size of BL 7.1, the largest baseline in ESLOC, is 1.0.

5 Figure 5.3 depicts the data in Table 5.6.

6 The lines represent least-squares linear fits to the four data points.

Figure 5.2
Normalized ESLOC Versus Facility Hours for Historical Aegis Baselines and ACBs

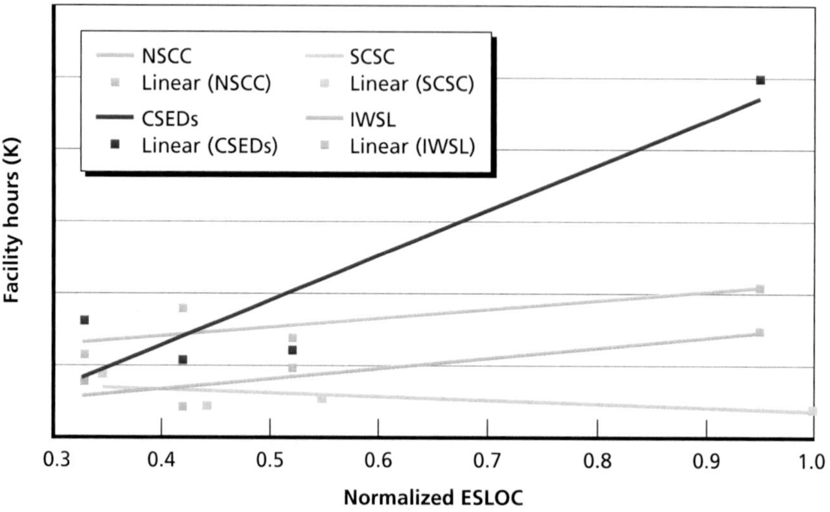

RAND *RR161-5.2*

But does the I&T effort correspond to calendar time? The calendar time required for I&T of historical baselines and ACBs has varied considerably—from 65 months for BL 7.1R to 29 months for BL 6.3. Historically, calendar time has not varied with the level of effort. For example, BL 7.1R and BL 7.1 have comparable I&T times but radically different facility hours. However, we understand that calendar time for I&T was extended for BL 7.1R due to the effects on shipyards from Hurricane Katrina in August 2005. If we correct this outlier by replacing the BL 7.1R I&T calendar time with the time given to ACB-08, we see that, in fact, calendar time would vary as expected with the level of effort and ESLOC.

Thus, historical evidence suggests that the Aegis development enterprise may be unable to integrate and test an upgrade in less than 2.5 years, and the upgrade frequency must be chosen to correspond with the size and complexity of the upgrade.

Trade-Offs Between Upgrade Size and Frequency

Conceptually, it would be infeasible to field large upgrades quickly. For example, there is no evidence that ACB-TI combinations of a similar size to BL 7.1 or ACB-12 could be integrated and tested even within a three-year window. Conversely, fielding small upgrades infrequently would be an inefficient use of the development enterprise because mature capabilities would be developed but not harvested. For example, the Aegis development enterprise has demonstrated that it can integrate and test ACB-08–sized upgrades in 31 months; if the necessary capability were available today, there would be no reason to wait four years for I&T.

Thus, the Navy's choice of ACB and TI upgrade frequency must consider an obvious trade-off: PEO Integrated Warfare Systems can either field relatively large upgrades less frequently or relatively small upgrades more frequently.

Interviews across the Aegis enterprise suggest that there are fixed costs to integrating and testing new capabilities that the Navy must pay regardless of the size of the upgrade. Examples include the cost of Combat System Ship Qualification Trials, certification testing, and operational testing to ensure the compatibility of the upgrade with the broader weapon system. There are also variable costs that depend on the size or complexity of the upgrade. Examples of variable costs that may increase with the size of the upgrade include the costs of regression testing and other developmental tests.

Smaller, more frequent upgrades will distribute capability improvements across the fleet more quickly, but the fixed costs of I&T will consume more of the fixed annual budgets. Larger, less frequent updates will distribute capability more slowly, but the fixed costs of I&T will consume a smaller amount of fixed annual budgets. Figure 5.3 illustrates this trade-off between upgrade frequency and size in a conceptual choice framework.

Some historical evidence supports the importance of distinguishing between fixed and variable I&T costs. Historical CSEDS, NSCC, and IWSL usage illustrates a positive relationship with ESLOC, as expected. SCSC demonstrates a weak negative relationship

Figure 5.3
Trade-Offs Between Aegis Upgrade Frequency and Size

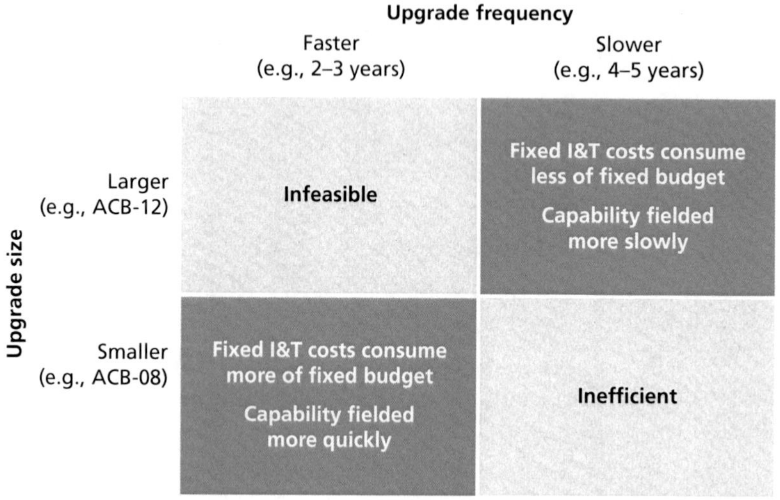

RAND *RR161-5.3*

between usage and ESLOC, which is likely an artifact of our data (see Figure 5.2).

What is more interesting to note is that the variable component appears to be a more significant driver of usage at CSEDS than at NSCC, IWSL, or SCSC; this is reflected in the more positive slope of the line representing a linear fit for CSEDS as compared with the others. Thus, in general terms, one might conclude from the data that activities at CSEDS are determined by the size of the upgrade, and activities at NSCC, IWSL, and SCSC are relatively fixed.

The available data limited our ability to estimate precisely variable and fixed costs and to identify specifically where these costs are introduced. We have only four historical data points (ACB-08, BL 7.1R, BL 7.1, and BL 6.3). Moreover, three of them are comparable in terms of ESLOC (ACB-08, BL 7.1, and BL 6.3); so in some sense, the data represent only two cases. Still, historical data provide evidence of the existence of these costs and rightly illustrate the trade-off that the Navy faces.

Trade-Offs for the Aegis Workforce and Facilities

To gain more insight into the trade-offs between larger, slower upgrades and smaller, faster ones, we estimated the demand on facilities and workforce under the IWS business model. Since the Navy has not specified the size of future ACBs, we assume that future ACBs and TIs resemble ACB-08 in the following ways:

- There is a fixed cost of integrating and testing future ACBs or TIs, which we model as 20 percent of the monthly usage or manpower of ACB-08.
- The marginal additional facility usage and workforce demand due to integrating and testing future ACBs (with or without a simultaneous TI) is modeled as 60 percent of the monthly usage or manpower of ACB-08.
- The marginal additional facility usage and workforce demand due to integrating and testing future TI (with or without a simultaneous ACB) is modeled as 20 percent of the monthly usage or manpower of ACB-08.

The reason to model the marginal effort of future ACBs (or TIs) as 60 percent (or 20 percent) of monthly usage is that ACB-08 included both computing hardware and software upgrades, so technically, ACB-08 is both a TI and an ACB in the terminology of the IWS model. In effect, we assume that 60 percent of the ACB-08 effort was devoted to software upgrades, 20 percent to computing hardware upgrades, and 20 percent reflected fixed costs.

A consequence of these assumptions is that simultaneous I&T of a future ACB and TI will be 100 percent of the monthly usage or manpower of ACB-08; if future ACB and TIs are staggered, monthly usage and manpower to integrate an ACB will be 80 percent of ACB-08 and 40 percent of ACB-08 to integrate a TI.

We further assume that the annual capacity (i.e., the maximum facility usage or manpower) of the Aegis enterprise is set to the average yearly usage or manpower level of FYs 2000–2010. This represents a fixed constraint on budget, manpower, and facility space.

Figures 5.4–5.11 show the results of our modeling. Figures 5.4–5.6 show projections of the amount of enterprise manpower that would be devoted to I&T. Figures 5.7–5.9 show projections of CSEDS facility usage during periods of I&T. The green lines correspond to usage due to technical insertions, and the blue lines correspond to usage due to advanced capability builds. Figures 5.10 and 5.11 show the aggregate effort over the projected decade across a range of ACB and TI intervals. In all cases, the vertical axis represents project effort (or usage) as an average annual effort/usage exhibited over the last decade.

Faster upgrade frequencies consume more facility hours and manpower, and leave less available manpower or facility hours for development. In other words, the fixed costs of I&T consume more of the total capacity of the enterprise. Offsetting ACBs and TIs has minimal impact on the aggregate level of effort or usage; however, Figures 5.5 and 5.8 show that offsetting the TI and ACB has the effect of level-loading the infrastructure.

Figure 5.4
Projected Aegis Enterprise Level of Effort Devoted to Periodic I&T (ACB interval = 4 years, TI interval = 4 years, no offset)

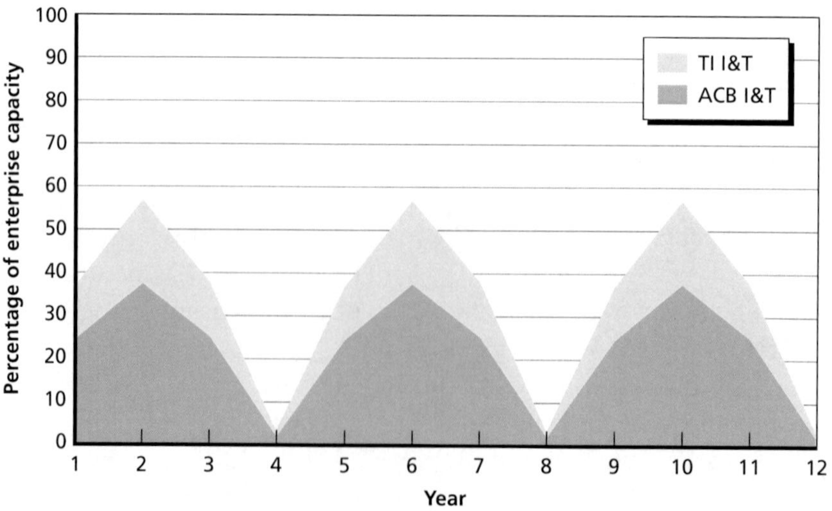

RAND *RR161-5.4*

Figure 5.5
Projected Aegis Enterprise Level of Effort Devoted to Periodic I&T (ACB interval = 4 years, TI interval = 4 years, 2-year offset)

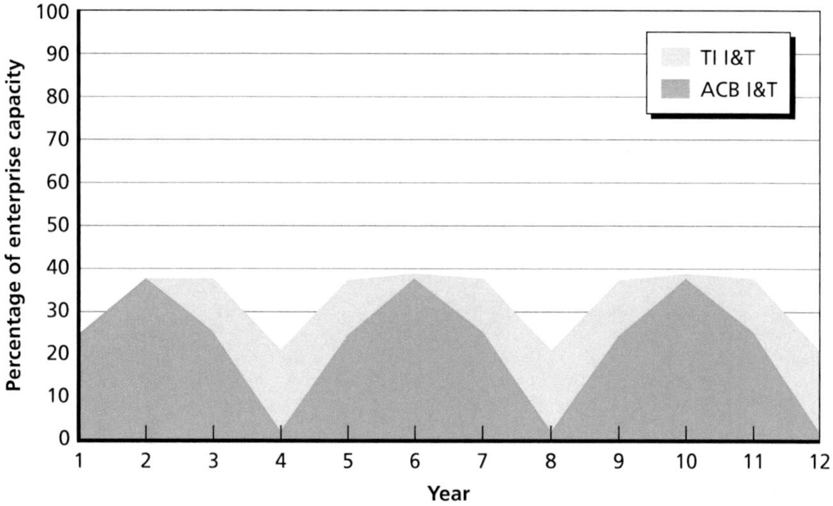

RAND *RR161-5.5*

Figure 5.6
Projected Aegis Enterprise Level of Effort Devoted to Periodic I&T (ACB interval = 2 years, TI interval = 4 years, no offset)

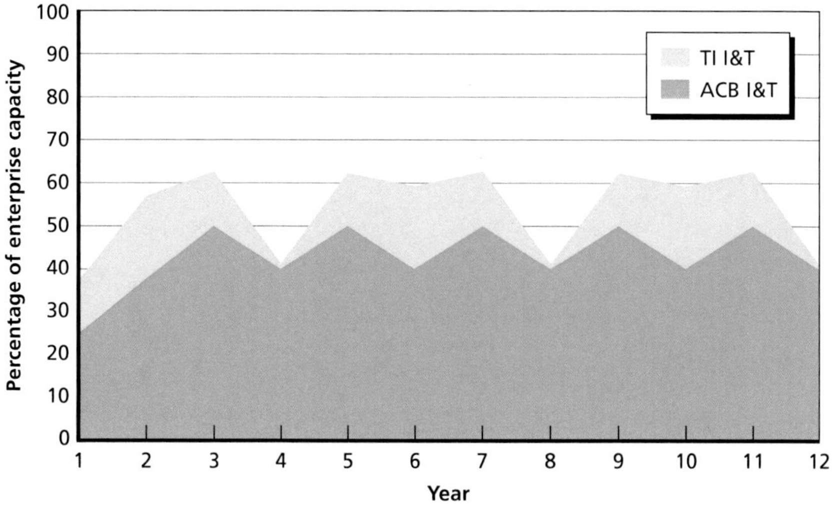

RAND *RR161-5.6*

Figure 5.7
Projected CSEDS Usage Devoted to Periodic I&T (ACB interval = 4 years, TI interval = 4 years, no offset)

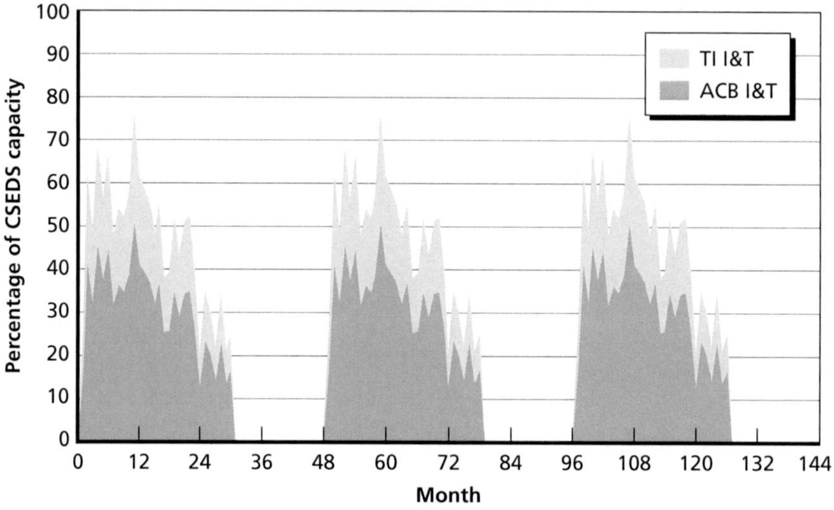

RAND *RR161-5.7*

Figure 5.8
Projected CSEDS Usage Devoted to Periodic I&T (ACB interval = 4 years, TI interval = 4 years, 2-year offset)

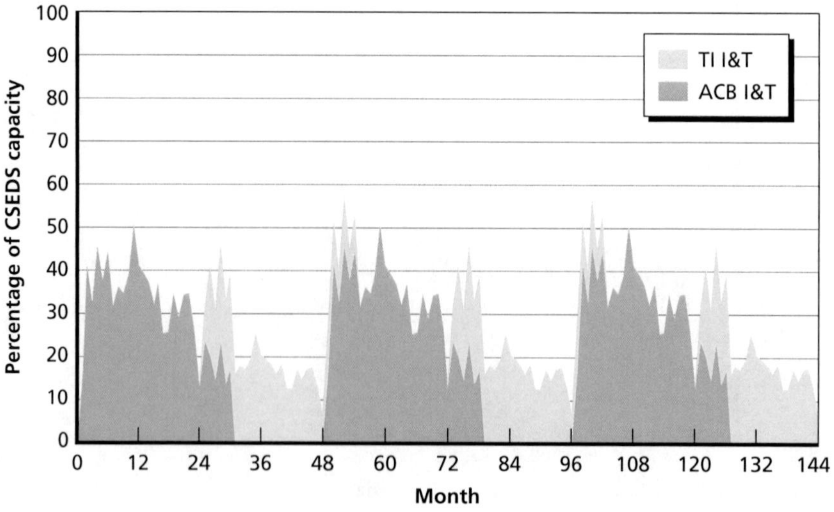

RAND *RR161-5.8*

Figure 5.9
Projected CSEDS Usage Devoted to Periodic I&T (ACB interval = 2 years, TI interval = 4 years, no offset)

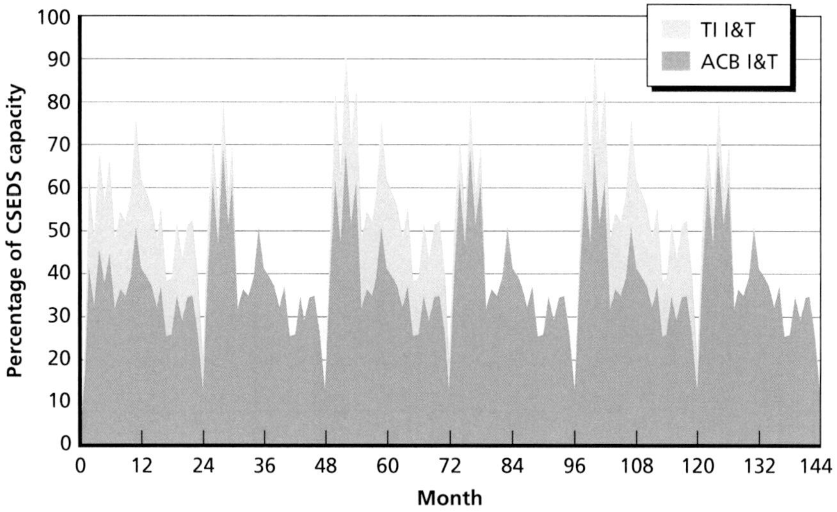

RAND *RR161-5.9*

Figure 5.10
Summary of Projected Aegis Enterprise Levels of Effort Devoted to Periodic I&T (all options)

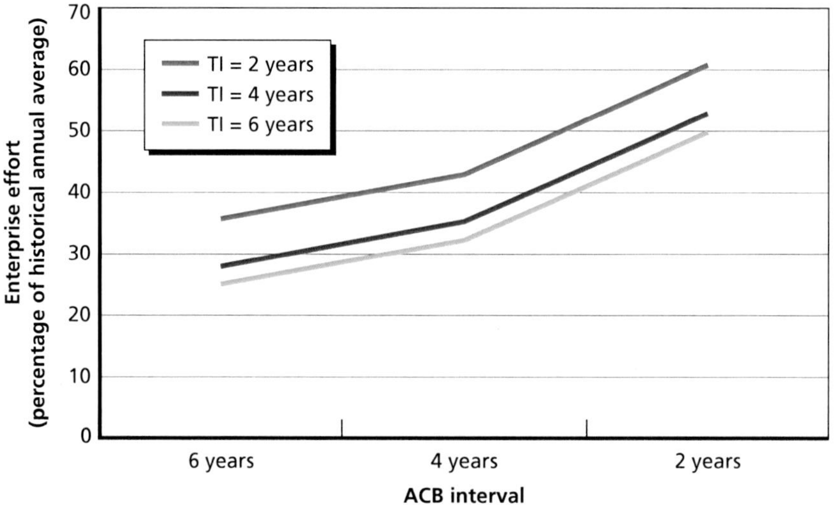

RAND *RR161-5.10*

Figure 5.11
Summary of Projected CSEDS Usage Devoted to Periodic I&T (all options)

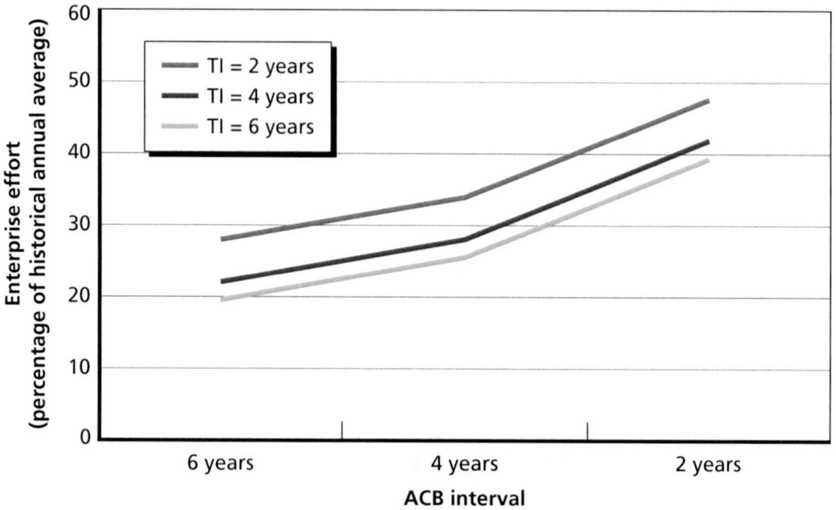

RAND *RR161-5.11*

Additional Considerations for TIs

TIs have the additional challenge of coordinating Aegis computing and network hardware upgrades with the commercial industry that provides the COTS equipment.

A faster pace for TIs means that the Navy will always deploy computing equipment at or near the state of the art. It also means that in-service issues can be resolved by the commercial IT industry; hardware can be addressed through the commercial marketplace. A slower pace for TIs may cause computing hardware to lag behind the state of the art, which may result in high costs to replace in-service hardware if replacement parts are not in production. Historically, the cost of COTS hardware follows a hockey stick pattern, whereby the cost of purchasing the most recent technology is greatest, the cost of purchasing one-generation-old technology is lowest, and the cost of purchasing legacy technology can be very high if the technology is out of production. This suggests that choosing a TI interval of around two years would minimize costs and allow the Navy to field technology that is just one or two generations behind the state of the art. These costs and benefits are in addition to those associated with the fixed and variable costs of I&T. Notably, this is the approach taken by the ARCI program.

While there are very good reasons to stay as close to the pace of the commercial world as possible, the Navy, as mentioned, has never integrated a baseline in less than 2.5 years, so we have no reason to expect that a two-year TI cycle is either feasible or necessary. In other words, the Navy is not currently using the increased capabilities of better hardware. In particular, the Navy is not presently leveraging available computing capacity (Miller, 2010, slide 8). Although this could change with new threats, radars, and algorithm technology, the procurement and I&T costs of faster upgrades may not come with capability benefits in the near term. Moreover, the Navy may be able to mitigate the in-service costs of slower upgrades by warehousing older computing hardware after TI upgrades and using the warehoused hardware for in-service support to the fleet. The idea is that the old COTS hardware from upgraded ships would be warehoused as a replacement pool used to service ships. This approach mitigates the cost of procuring legacy hardware upgraded at slower intervals and happens to be the approach followed by ARCI.

Summary

The IWS business model to integrate and test new hardware and software upgrades every four years is consistent with legacy efforts. Of the four most recent development efforts, BL 6.3 was integrated and tested the fastest, and that took 2.5 calendar years; ACB-08 was integrated and tested in 31 months.

As the Navy considers alternative upgrade intervals, it must balance upgrade size and frequency. Smaller, more frequent upgrades will distribute capability improvements across the fleet more quickly, but the fixed costs of I&T will consume more of the fixed IWS budgets. Larger, less frequent updates will distribute capability more slowly, but the fixed costs of I&T will consume less of the fixed budgets. Our analysis of available data confirms that this trade-off is real.

The choice of TI interval requires the additional consideration of coordinating computing and networking hardware upgrades with the industry that produces COTS equipment. ARCI has followed an approach of upgrading hardware roughly every two years, which allows it to minimize procurement and in-service costs while leveraging recent, if not state-of-the-art, COTS equipment. Unfortunately, integrating new hardware in two years has been historically infeasible for Aegis. However, the Navy may be able to mitigate the in-service cost of slower drumbeats by warehousing retired computing hardware. Also, to date, it has not required the computing capacity provided by faster upgrades.

CHAPTER SIX

Risks

The IWS business model calls for changes in the way weapon systems are developed. These changes in process may affect the people and facilities involved with developing weapon systems. In some cases, the necessary changes may render some implementation options infeasible, since they may impose too great a demand on the Navy's personnel and facilities. In other cases, the changes may be feasible but introduce risks that could, in the end, be passed along to the warfighter.

This chapter discusses the risks inherent in the IWS business model and describes ways the Navy could mitigate them. Our list of risks is not exhaustive; rather, it represents a set of issues that arose over the course of our research. Each issue was confirmed as a risk through conversations with U.S. Navy officials in the Aegis, SSDS, and ARCI programs and with industry developers.

Sources of Risk

The Switch to a Completely New Business Model May Entail Unanticipated Difficulties

The fundamental differences between the IWS model and the historical (legacy) approach to developing and fielding weapon systems are perhaps the main sources of risk in the IWS plan. These differences were discussed in greater detail in Chapters Two and Four of this report. The Navy should make no mistake: The IWS business model calls for an entirely new approach to developing and acquiring weapon systems that could have consequences that this report or the Navy itself may be unable to anticipate. That is, beyond the anticipated consequences for facility usage, manpower, roles and responsibilities, and so on, the plan may have influential second- and third-order consequences. Such consequences may be difficult to predict, given the size and complexity of the Aegis program.

The Vested Interests of Stakeholders in Legacy Process May Make Implementing a New Business Model More Difficult

As discussed in Chapter Three, AWS development relies on a large enterprise of independently managed industry and government organizations. Four government organizations play a significant role—NSWC Dahlgren, NSWC Port Hueneme, Aegis TECHREP, and PEO Integrated Warfare Systems—and our data suggest that, in some cases, the government level of effort is comparable or larger than that of the industry organizations. For example, we estimate that the government man-years for core Aegis baselines in the most active periods of I&T was at least 60 percent of the total enterprise man-years in FYs 2005–2010.

During the transition period of that decade, cross-organization groups formed to help manage the distributed Aegis enterprise. The Fleet Change Review Board and the Senior Change Control Board, for example, include representatives from across government organizations. However, funding (and therefore incentives) remains structured along organizational lines, and organizational efficiencies in general do not translate to enterprise efficiencies.

To be clear, the effective collaboration between industry and government organizations reflects the celebrated success of the Aegis program, but it is also a source of risk. By necessity, the IWS business model will be implemented from the top down and must be embraced by an enterprise that is heavily invested in the legacy approach. While any change may be greeted with some resistance, the magnitude of this particular change may conspire with historic interests to create obstacles to its implementation. Our interviews with Navy officials familiar with the ARCI experience confirm that this source of risk should not be ignored.

More concretely, all four government organizations have historically played some role in I&T. The IWS business model calls for a highly regimented I&T timeline that is unprecedented for the AWS, and failure to adhere closely to this timeline will have cascading effects on the capabilities that are distributed to the fleet. The risk of departing from the timeline may increase if too many players are allowed to make decisions that influence it. Furthermore, the plan calls for fielding new capabilities every four years, and there may be pressure to increase the fielding rate as new threats and technologies emerge. A high level of government involvement may increase the fixed costs of I&T; following the logic outlined in Chapter Five, this may render faster drumbeats infeasible.

The Complexity of Managing the CSL Adds Risk

One of the most significant changes is that development activity will be managed through a CSL. As discussed earlier, this means that different ACB-TI combinations will be managed from a single, shared software library. This approach differs significantly from the legacy approach, in which software for new baselines began as a "clone" of older baseline software and was then managed separately. The benefit of using a CSL is that improvements and fixes for one ACB-TI combination can naturally propagate to others that rely on the same software. However, implementing a CSL represents a significant management challenge. The challenge is managing the library so that different development activities that require simultaneous access to the same software component do not interfere with one another. This will likely require the expertise of individuals who have managed projects of similar scale and complexity.

Figure 6.1 illustrates the potential consequences of *not* using a CSL. The blue bars represent the number of residual computer program change request (CPCRs) from the conclusion of the ACB-08/BL 8 development effort; CPCRs are broken out by CPCR risk category, which ranges from R1 to R4. The red bar indicates the number of those ACB-08/BL 8 requests that were recommended to be fixed in ACB-12, broken out across the same risk categories. The discrepancy indicates that many of the problems that were discovered while inte-

Figure 6.1
Distribution of Open ACB-08 CPCRs Across U.S. Navy–Reported Risk Category

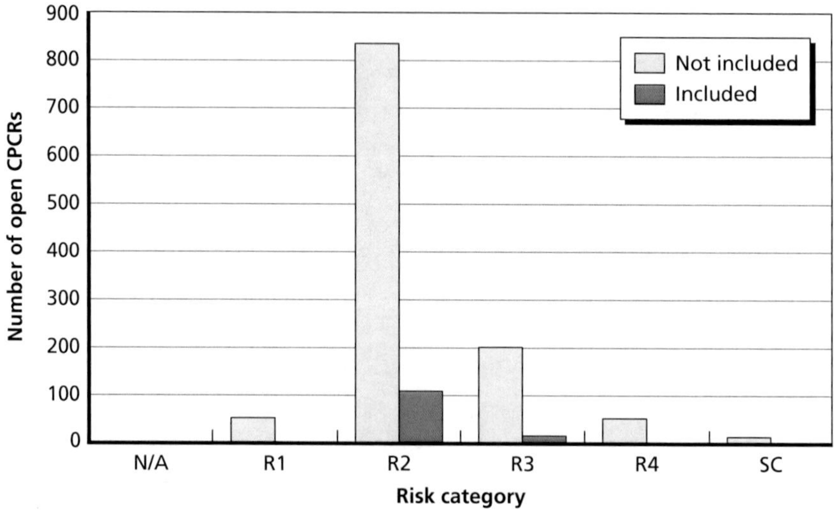

RAND *RR161-6.1*

grating and testing ACB-08/BL 8 will not be fixed in ACB-12. Historically, the ACB-08/BL 8 CPCRs may have been addressed during the support phase of ACB-08/BL 8, even if they are not incorporated into ACB-12/BL 9. In general, without a CSL, a single CPCR would need to be fixed multiple times, once for each affected ACB. With a CSL, a CPCR would need to be addressed only one time because the fix would propagate through the shared library.

The Navy expects to develop and field new capabilities frequently and widely across the fleet. In this environment, the fleet is likely to increase its demand for changes (because more ships will receive a given upgrade). The increased pace of development also means that, if CPCRs are not addressed at a rate comparable to that at which they are introduced, the overflow of CPCRs could compound, yielding unreliable or dangerous software. Without a CSL, the consequences of having CPCRs overflow could increase to the point that it may be impossible to implement the model without one. Since the time this research was conducted, the Navy reports progress in fielding a CSL. As part of the

Accelerated Mid-Term Interoperablity Improvement Project (AMIIP), software fixes resulting from CPCRs for Aegis BL 6, BL 7, and BL 8 ships have been incorporated into the CSL and inherited by ACB-12/ BL 9 without requiring a separate development effort.

Conversations with software architects and a literature review suggest that modern software engineering processes and practices routinely use concepts similar to the CSL. For example, the SSDS program has adopted a CSL-type approach (albeit on a smaller scale). In other words, the IWS model is not without precedent.

The Navy and the Missile Defense Agency's BMD Program Poses the Risk of Competition for Limited Resources

In the future dictated by the IWS business model, the Navy and MDA's BMD program will operate under shared resource constraints. If not carefully managed, the result could be a reduction in overall capability for both programs. In this section, we consider four obvious areas of potential competition for resources.

First, the Navy and the BMD program may compete for manpower and for space and time at I&T facilities. Presently, the Navy and the BMD program use common facilities and personnel to develop and support Aegis and Aegis-BMD. Lockheed Martin develops radar and weapon system capability for both groups, and NSWC Dahlgren and NSWC Port Hueneme have roles in both programs. Moreover, both use CSEDS, SCSC, NSCC, and IWSL facilities. The capacity of these facilities is likely to remain fixed. This introduces the possibility that the two programs will compete for facility capacity to support their independent development activities. BMD demands on Aegis facilities have increased between FY 2000 and FY 2010, as shown in Figures 3.4 and 3.5, and the IWS model clearly anticipates increasing U.S. Navy demand. Thus, the data also foreshadow competition for facility time.

Second, the Navy and the BMD program may compete for shares of individual ACBs. The IWS plan calls for integrating Aegis and Aegis-BMD systems at a software level, in part to enable programs to share common software components. The size of ACBs (measured, for example, in ESLOC) will be relatively fixed to ensure that new capabilities can be integrated and tested in the proposed timelines. As a result, the two programs may compete to field their priority capabilities in a given ACB. For example, one can envision scenarios in which both groups have high-priority capabilities (or CPCRs to address) that they wish to field, yet in which it is infeasible to integrate and test both capability sets in the planned timeline.

Third, the Navy and the BMD program may compete for computing resources in a given TI. PEO Integrated Warfare Systems reports that the computing capacity available to Aegis and Aegis-BMD is underutilized at present. But future threats, new radars, and new advanced signal processing algorithms may increase demands on the microprocessors. As a result, the two programs may compete for computing capacity.

Finally, during development, the Navy and the BMD program may compete for access to software components in the CSL. As mentioned, the IWS model calls for programs of record, software fixes, and other rapid capability insertions to be managed simultaneously in a CSL. Ideally, the software will be designed so that independent development activities are unlikely to require simultaneous access to the same components. But in the short term, before the software is fully compartmentalized, overlap may be likely, and in the long term, some overlap may be unavoidable. This may delay development or even cause some Navy or BMD capabilities to be fielded later than originally planned.

Investments in Product-Line Development and Capability May Compete for Limited Resources

We described the objectives of the IWS business model in Chapter Two. The objective of leveraging capability development across weapon systems is fundamentally independent in the sense that its achievement is not expected to contribute to meeting the other objectives. Moreover, it is not expected to produce direct improvements in warfighting capability, and it does not contribute to implementing or managing the necessary CSL.

Leveraging the overlap between systems, however, will likely be costly. ACB-12 brings the first addition to the CAL—the Joint Track Manager—at an estimated cost of $100 million. If the cost of adding other weapon system components to the CAL proves similar, the Navy would be wise to consider putting those monies directly into capability improvements. Thus, there is a risk that investments in the so-called product-line approach to development will divert resources from efforts that bring capability to the warfighter more directly.

IWS Business Models Expose the Aegis Fleet to New Risks as a Result of Funding Instability

As with any program, the success of the Aegis depends on a stable line of funding. In the legacy model, shifts in funding would primarily affect the timelines for new construction and the midlife upgrade timelines. The long timelines associated with new construction and midlife upgrades are such that year-to-year funding instability may be easily managed by simply managing funds for an appropriate number of budget cycles. However, in the planned model, funding instability may introduce more subtle changes in fleet dynamics that can propagate through the fleet for years to come. Specifically, there are three risks.

The first risk is funding for Aegis modernization. If insufficient funding is available to modernize ships that currently host legacy Aegis baselines, those ships will not enter of the pool of ships available for periodic ACB or TI upgrades. The larger the pool of unmodernized ships, the greater the fleet is exposed to obsolesce risks inherent to the legacy business model. Moreover, a modernized fleet is a prerequisite for obtaining the principle benefit of the new business model—the ability to spread IWS research, development, testing, and evaluation (RDTE) investments across the entire fleet. In short, reductions in the modernization rate would decrease the percentage of the fleet that can receive a new capability, increase the average age of software and hardware in the fleet, and potentially increase the number of different configurations deployed.

A second funding-related risk has to do with funding for fielding ACB and TI on previously modernized ships. If funding is not available for ACB and TI fielding, there is the potential for obsolescence (as computing hardware is required to extend the life of the Aegis software). And again, ACB developments would not propagate across the fleet. This would not allow the costs of IWS ACB development to be spread across the fleet. A reduction in the upgrade rates would increase the average age of software and hardware in the fleet and potentially increase the number of different configurations deployed.

The final funding-related risk is to ACB and TI development. If funding is not available to develop new capabilities, then there will be no capabilities to distribute across the modernized fleet. In this case, ships could receive upgrades only to fix CPCRs or make other minor changes.

Strategies to Mitigate Risk

There are various strategies the Navy could pursue to mitigate these risks. In this section, we consider a few approaches that set the stage for our recommendations.

Make Capital Investments in the CSL and Software "Componentization"

To mitigate the complexity of managing the CSL, the Navy should consider directly investing resources in the infrastructure and people required to manage it. These investments should be devoted to ensuring that the software tools necessary to implement the CSL are available and that the appropriate personnel are hired or trained to manage it. These investments should not be tied to a specific development effort, ACB, or TI; rather, they should be considered a capital investment in infrastructure, since the CSL is best thought of as an infrastructure for future developments.

In addition, the Navy should consider investing in componentizing software so that the CSL can be fully leveraged. In this context, *componentization* refers to structuring software into separate pieces so that changes to any one component are unlikely to require simultaneous changes to other components. Componentization facilitates the implementation of a CSL by decreasing the chance that independent development activities will require simultaneous access to the same component. Indeed, some level of componentization is likely a prerequisite for a CSL.

PEO Integrated Warfare Systems reports that once ACB-12 is completed, the Navy will be on the path of having componentized Aegis software, but more work will likely be necessary to componentize the software that Aegis shares with Aegis-BMD. Further, at the time of this writing, the Navy has made investments toward implementing a CSL. Further investments in componentizing software may mitigate the risks associated with managing the CSL.

Streamline Government Involvement in I&T

In general, streamlining I&T processes will increase the likelihood that planned capability builds and technology insertions will enter the fleet in accordance with the regimented timetable anticipated by the IWS model. As we have discussed, the government is heavily involved in I&T. Thus, a natural way to increase efficiency is by streamlining the government's role.

Streamlining the government's role in I&T does not necessarily mean streamlining its overall role in developing and supporting the AWS. The government should maintain its necessary roles in establishing requirements, conducting at-sea test trials, and fielding and providing in-service support.

Notably, successful implementation of a CSL should improve the efficiency of the I&T process. Further, operating with a CSL would mean that software fixes would be captured and propagated forward. Each ACB or TI upgrade should increase the stability of the code. The combination of sequential I&T and improvements to the code base using the CSL may allow government organizations to pare down their role in this process, potentially to a significant degree.

Enforce Requirements Discipline

A recurring difficulty in any acquisition program is managing the requirements process. In the case of Aegis, there is a desire to include as many requirements as possible in a given baseline. Currently, ships that receive a certain baseline are not expected to be upgraded until their midlife upgrade. As a result, there is pressure to include capabilities that may not be mature enough to meet the development timeline. In addition, there is a tendency to include capabilities, often based on fleet feedback, late in the development process. This combination of behaviors makes testing difficult and often leads to the inclusion of software errors.

The IWS model offers relief from these pressures. Capabilities are delivered to the fleet on a periodic basis. As a result, if a piece of technology is not mature enough for inclusion in a particular update, or if the requirement for a capability is not identified early in the ACB process, it can be delayed until the next update with minimal effect on the fleet. However, due to the sequential I&T requirement of the IWS model, there is less ability to delay the introduction of the next upgrade. Upgrades must be available on schedule because of the large number of installations required annually. Delays cause ships to miss their installation and can derail the entire process.

Stagger TIs and ACBs

Another approach to mitigating the risks of not meeting planned I&T timelines is to stagger TIs and ACBs so that hardware and software are never changed simultaneously. The idea is that, in one round, a new ACB would be introduced on the most recently certified TI; in the next round, a new TI would be introduced on the most recently certified ACB. A rival approach would be to change ACBs and TIs simultaneously. The advantage of the former approach is that it limits the complexity of the capability entering the I&T phase. Together with requirements discipline that limits the size of ACBs, this strategy may help reduce the likelihood that an ACB-TI combination cannot be integrated and tested in time to meet planned fielding dates. Assuming that software fixes are included in both ACB and TI upgrades, staggered upgrades would provide twice the number of opportunities to improve code stability. Implementing and resourcing a CSL, combined with dedicating resources for software fixes, would improve stability and performance.

Delay Investments in Product-Line Development Until the Transition to a CSL Is Successful or a Cost-Benefit Analysis Is Conducted

Delaying investments in product-line development would eliminate the risk of resources being diverted from capability improvements to fund the product-line approach. This delay may be warranted because, as we have discussed, investments in the CAL have been and are expected to continue to be expensive, and they are not expected to provide direct improvements to Aegis warfighting capability. The Navy could reconsider investments in the CAL and a product-line approach after it has successfully transitioned to managing development via a CSL—another significant source of risk.

Harvest Tactical Lessons Learned from ARCI and SSDS

The IWS business model resembles, in some respects, the SSDS program and the approach the submarine community takes in the ARCI program. This report has attempted to harvest lessons learned from those programs (summarized in Chapter Seven), but we expect that there is more to be learned, especially at a tactical level. For example, there may be lessons that Lockheed Martin's Aegis operations can harvest from its work on ARCI about software engineering using a CSL. The Navy should not assume that disparate groups within Lockheed Martin are communicating with one another. For another example, when it was implemented, ARCI represented a dramatic restructuring of the way government warfare centers supported combat system development; the Navy could use that experience to assess the costs and benefits of alternative uses for its warfare centers.

Summary

There are several sources of risk in the IWS business model. The most fundamental risk, which should not be underestimated, stems from the fact that the model is very unlike the legacy approach to developing and fielding Aegis capability upgrades. The experience of ARCI and SSDS provides some lessons learned, but Aegis is a fundamentally different program, and the Navy should expect it to have a unique level of complexity. Thus, the Navy should proceed slowly when implementing this plan and be prepared to harvest its own lessons as it proceeds to develop upgrades with a CSL and field these upgardes to modernized ships.

Another significant source of risk in the IWS model is the fact that multiple government stakeholders have a vested interest in the legacy business model. Other important risks include the complexity of managing a CSL; the possibility that the Navy and BMD will compete for limited talent, facility time, and access to the CSL; and the diversion of resources to the CAL from direct capability improvements. The Navy can mitigate some of these risks by making capital investments in the CSL, through software componentization, by delaying investments in product-line development until transition to the CSL is successful, by streamlining government involvement in I&T, by enforcing requirements discipline, by staggering ACBs and TIs, and by harvesting lessons learned from ARCI and SSDS.

CHAPTER SEVEN

Lessons Learned from ARCI and SSDS

The submarine community transitioned its fleet to an OA-based computing architecture in the mid-1990s with the introduction of the ARCI program. It has subsequently added the combat system and above-water sensors to its OA model. Currently, the entire attack (SSN) and ballistic missile (SSBN) submarine fleets operate with OA sonar, combat, and above-water sensor systems. The Navy deploys SSDS on its carrier and amphibious ships. SSDS, which focuses on own-ship protection, was developed and continues to evolve using an OA approach. In this chapter, we discuss similarities and differences in these OA systems relative to Aegis and look for potential lessons learned.

ARCI Lessons Learned

In the mid-1990s, the U.S. submarine force experienced degradation in its acoustic advantage due to the quieting of Russian nuclear submarines and the widespread introduction of extremely quiet diesel-electric submarines. While continuing to improve its own noise signatures, the submarine community needed to improve its sonar capabilities. The standard submarine sonar systems in the 1990s, the BQQ-5 on SSNs and the BQQ-6 on SSBNs, operated with proprietary software running on MILSPEC computers.[1] Upgrading those systems in response to the evolving submarine threat was neither technically nor fiscally

[1] The standard shipboard computer was the Sperry UYK-7, which was designed to demanding performance and ruggedness specifications.

feasible. As a result, the Navy's submarine fleet turned to OA to exploit commercial computing technology and allow for more frequent software upgrades.

The submarine OA program covers three areas: acoustics, combat systems, and above-water sensors. The acoustic and combat system programs are directly relevant to the proposed Aegis OA efforts. The ARCI program delivers software updates (advanced processing builds [APBs]) and hardware updates (TIs) every two years. The APB and TI developments are offset, such that APBs are delivered in odd years and TIs in even years. Each TI also includes computer maintenance repairs that are transparent to the fleet. Individual submarines receive every other TI upgrade and, depending on their deployment schedule, every APB.

The current practice of fielding APBs and TIs every two years, with the updates offset into odd and even years, is a byproduct of more than ten years of fleet experience. Initially, the ARCI program developed and fielded a new APB every year. However, fleet concerns about training, tactics, and procedures resulted in the current two-year drumbeat. Offsetting the software and hardware upgrades was also a deliberate decision. The offset means that software upgrades are developed and applied to mature hardware and vice versa. Thus, any potential development issues can be isolated to either the software or hardware. In addition, it forces the software to operate across multiple hardware configurations, maintaining the OA design philosophy. Finally, it should be noted that, as the ARCI model expanded from the *Los Angeles*–class to the *Ohio-*, *Seawolf-*, and *Virgina*-class submarines, the engineering development effort required for a given upgrade increased. The program faced the same difficulty when it incorporated the combat system and above-water sensors. These increased challenges of implementing software and hardware upgrades across multiple ship classes and a large suite of applications and sensors are shared by Aegis.

Insights from the ARCI Experience

ARCI and Aegis OA efforts both feature periodic software and hardware upgrades made possible by separating application software from commercial hardware via COTS-based middleware. The submarine community's ARCI experience may provide insights into the periodicity of updates, development and testing of the OA system, programmatic challenges, and fleet impact. There are obvious technical differences between detecting, tracking, and engaging submerged targets and supersonic missiles that must be considered in any comparison.

The ARCI experience demonstrates that fleet performance improvements make the difficulties of transition worthwhile. Prior to ARCI's development, the Navy's submarine fleet was losing its acoustic advantage. The incorporation of commercial computer hardware immediately increased the available computing power, allowing the incorporation of sophisticated acoustic algorithms into the submarine sonar system (Jacobus, Yan, and Barrett, 2002). These algorithms, aided by sophisticated display technology, improved sonar operators' time to initial contact and contact hold time by 45 percent and 25 percent, respectively (Zarnich, 2006). Subsequent APBs have improved operator performance in time to initial contact by an additional 80 percent and contact hold time by an additional 13 percent. There have been improvements in system reliability as well, though the pace of these improvements has not been as startling.

The choices made regarding ARCI TI and APB upgrade periodicity offer potential lessons for Aegis. The ARCI program develops a new TI every two years. This two-year TI periodicity is driven by the computing capacity required to support APB development and allows ARCI to maintain pace with the computing industry. The commercial computing industry evolves on an 18- to 24-month cycle (Kerr, 2006). Maintaining pace with commercial industry allows components to be replaced prior to obsolescence. Because the commercial computer industry does not support out-of-production components, a TI schedule matched to the industry timetable minimizes procurement and in-service support costs. As the installed computer hardware gets older, the purchase and in-service support costs increase, driving the program toward faster upgrades. On the other hand, installing new hardware across the fleet is expensive, driving the program toward slower upgrades. In the case of ARCI, the combination of computing capacity requirements and the pressure imposed by fielding costs resulted in a two-year TI upgrade cycle.

The periodic nature of APB/TI upgrades in the ARCI model requires a consistent, stable funding level. Technology is not chosen for inclusion in an APB until it is mature. This necessitates a development effort that is capable of supplying sufficient technology to support the APB process. Individual APBs simply harvest technologies that have matured through ARCI development efforts and then integrate them into the combat system. The development and installation of TIs also requires a stable source of funding.

OA increases the diversification of source code suppliers. Under the OA model, the modular software code opens the software development process to competition. ARCI competes individual software modules among large prime contractors, U.S. Navy labs, universities, and small businesses. In particular, the OA software design has hastened the ability of small businesses to participate. Small business participation offers advantages such as increased innovation and cost reductions. The broader set of participants is particularly useful in the ARCI program due to the primacy of acoustic algorithms. Competition among universities, U.S. Navy labs, and industry to produce the next generation of acoustic algorithms has directly benefited the Navy's submarine fleet.

Limitations in the ARCI Model for Aegis

ARCI provides an excellent model for the Aegis program as the latter transitions to OA. However, there are limitations in applying those lessons to Aegis. ARCI is primarily a system that collects various acoustic inputs and analyzes the data by means of sophisticated algorithms to track and eventually engage with torpedoes. The speed and times involved are dramatically different for ARCI and Aegis. The surface or subsurface targets that ARCI engages are moving at speeds below 50 knots, and the difficulty rests primarily in resolving the ambiguities inherent in sonar tracking. Aegis manages engagements between supersonic missiles, implying a significantly shorter timeline. In addition, the acoustic environment of the submarine limits its interactions with other platforms. Aegis is part of a network of surface and aviation systems that share contacts and information. As a result, Aegis has to interact with numerous systems, each with its own modernization rate with which Aegis must maintain fidelity.

ARCI also has advantages over Aegis in its testing regime as it integrates each APB and TI. The ARCI program can run acoustic testing in the laboratory against real-world acoustic environments by using actual sonar recordings. At-sea testing of ARCI is also simpler. Torpedo testing is relatively inexpensive (the torpedoes are reusable) and can be done quickly. Aegis testing requires significant infrastructure, including missile ranges and expensive targets.

Finally, Aegis faces more severe organizational challenges relative to ARCI. The Navy's submarine fleet has more control over the management of individual upgrades. Decisions about what to include are all internal to the submarine community. Aegis, however, has two major organizations that provide input to capabilities. Each Aegis ACB will include improvements to the Navy's air and missile defense needs, as well as BMD enhancements. The Navy and the MDA must coordinate to determine which improvements will be incorporated in each upgrade.

Lessons Learned from SSDS

The SSDS combat system provides ship self-defense capabilities against anti-ship cruise missiles for aircraft carriers and amphibious ships. It integrates existing stand-alone sensors and anti-air weapon systems to provide an automated detect-to-engage capability against low-flying, high-speed anti-ship cruise missiles in the littoral environment. SSDS design emphasizes physically distributed nondevelopmental items, commercial standards, and computer program reuse in an OA computer network.

Although versions of SSDS are "released" in ACB updates, SSDS employs a process of never-ending debugging and development. SSDS-equipped ships typically receive updated SSDS software, regardless of where their refits fall in the ACB cycle. Updates to SSDS can be performed during maintenance periods, or updated software can be delivered to ships pierside. Because of SSDS's software architecture, updates to the system do not include an entire new version of the software. Its modularized design allows each component of SSDS to stand alone in the software architecture. Only software components that have been updated need be sent to the ship.

Insights from the SSDS Experience

SSDS was developed in a modern computing environment with modular software, COTS hardware requirements, and a single source library (SSL). SSDS is capable of easy updates and nonintrusive code additions because of the SSL. Through the SSL, software fixes in SSDS are tracked, incorporated into the master version, and distributed to the entire fleet. The SSL is functionally equivalent to the CSL being instituted in the Aegis OA effort.

SSDS has the ability to function on numerous hardware configurations across multiple classes of ships because programmers know what hardware exists on each platform and conduct installations accordingly. SSDS is not written specifically for a single computing hardware configuration. Although SSDS does not control all the systems it links, strict computing hardware rules are followed for ships running SSDS. These configurations allow the single version of SSDS to run on multiple ships and ship classes, all of which have different hardware configurations.

SSDS certification authority currently resides with Naval Sea Systems Command. SSDS certification appears to be somewhat simplified compared to that of Aegis. Small computing hardware changes at the component level, which require recertification within the Aegis program, do not have the same repercussions for SSDS. A lesson that Aegis can take away from SSDS is that small hardware changes do not necessarily need recertification. SSDS has shown that these sorts of changes at the component level that do not change functionality do not require a full recertification process. This saves time, money, and manpower and allows these resources to be directed toward the bigger-picture items.

CHAPTER EIGHT

Conclusions and Recommendations

As PEO Integrated Warfare Systems implements its new business model, it must carefully balance a range of associated costs, benefits, and risks. The OA Aegis system will be installed, beginning with ACB-12, on destroyers and cruisers already in the U.S. Navy's surface fleet. Risks inherent in this implementation will affect both today's fleet and the future fleet. For this reason, we propose an implementation approach for the IWS business model that minimizes potential risk but incorporates significant benefits to the fleet. It should be noted that one of the strengths of the proposed model is its future flexibility. As the Aegis technical community becomes proficient in developing, integrating, and installing ACB/TI upgrades, we fully expect the timing of this program to change in response.

RAND's ACB/TI Proposal

We agree with the IWS plan to field ACB and TI upgrades on a four-year drumbeat. In our proposed implementation approach, every ACB and TI upgrade is installed on every Aegis ship over each four-year period. Further, the ACB and TI upgrades are offset by two years. Figure 8.1 illustrates this proposed approach. In addition to new computer hardware, TI upgrades include both software fixes in response to CPCRs identified during the preceding ACB and modifications to the AWS as required to support ACS upgrades. For example, TI-18 would include software fixes identified by the fleet operating with the ACB-16 upgrades.

Figure 8.1
Proposed ACB/TI Implementation

<table>
<tr><th></th><th>2013</th><th>2014</th><th>2015</th><th>2016</th><th>2017</th><th>2018</th><th>2019</th><th>2020</th><th>2021</th><th>2022</th><th>2023</th><th>2024</th></tr>
<tr><td>ACB</td><td colspan="4">ACB-12</td><td colspan="4">ACB-16</td><td colspan="4">ACB-20</td></tr>
<tr><td>TI</td><td colspan="2">TI-12</td><td colspan="4">TI-14</td><td colspan="4">TI-18</td><td colspan="2">TI-22</td></tr>
<tr><td>Ship #1</td><td colspan="4">ACB-12/TI-12</td><td colspan="4">ACB-16/TI-14</td><td colspan="4">ACB-20/TI-18</td></tr>
<tr><td>Ship #2</td><td colspan="2">Legacy</td><td colspan="4">ACB-12/TI-14</td><td colspan="4">ACB-16/TI-18</td><td colspan="2">ACB-20/TI-22</td></tr>
</table>

RAND *RR161-8.1*

ACBs incorporate major capability enhancements and, as such, require significant I&T. Further, because they change the system's functionality, they have the biggest impact on the ship's operator. We recommend a four-year ACB drumbeat to balance the desire to deploy new capabilities with the risk of both compressed I&T times and the disruption to the ship's operations. The I&T requirements encourage a slower drumbeat, while the desire to deploy capabilities motivates a faster drumbeat. In the Aegis case, historical development efforts indicate that a two-year ACB drumbeat is infeasible. Further, a two-year drumbeat devotes a prohibitively large fraction of Aegis technical resources to I&T. This constrains development efforts that are critical to providing mature technology for subsequent ACBs. Finally, the planned capabilities in the Aegis technology roadmap do not easily break down into short, two-year cycles. BMD 5.0 and the planned Air and Missile Defense Radar require protracted I&T activities that do not easily fit into a fast drumbeat. To quickly deploy new capabilities to the fleet, our proposed approach involves installing each ACB on every ship—as opposed to every other upgrade, as in the ARCI model. This spreads I&T costs over a larger user base and maximizes the deployment of new capabilities to the fleet.

We recommend a four-year drumbeat for TIs, offset from the ACB upgrades by two years. Each TI upgrade includes both the commercial computing hardware and software fixes identified. Choosing a TI drumbeat requires PEO Integrated Warfare Systems to balance fidelity with the pace of the commercial computer industry, maintaining sufficient computing resources to support the capabilities installed with individual ACBs and minimizing costs. The I&T burden for individual TI upgrades is substantially lower than for ACBs. The commercial computer industry operates on an 18- to 24-month modernization schedule. Diverting from that schedule potentially increases in-service support costs. The TI drumbeat must also support the computing power required by the ACB capabilities. Normally, this would motivate a faster TI drumbeat. However, the Aegis upgrades incorporated in ACB-12 do not require the additional computing power offered by the switch to commercial hardware. The additional computing power in the TI-12 upgrade provides a hedge against future ACB requirements. Finally, installing new commercial hardware on the number of ships called for in the IWS business model is expensive. A four-year TI drumbeat minimizes the potential risks inherent in deviating from the commercial standard. Including software fixes in each TI upgrade doubles the opportunities to improve the stability of the Aegis code and to respond to operator-identified issues.

Offsetting four-year ACB and TI cycles balances deployed capabilities and development risk. Offsetting upgrade cycles has three advantages. First, software and hardware development efforts are isolated. Upgraded software, in the form of an ACB, is installed on mature hardware. Upgraded hardware, in the form of a TI, is installed on mature software. This allows system issues to be quickly isolated to either software or hardware. It has the additional advantage of reinforcing the separation between hardware and software. Individual ACB upgrades must operate on two TI upgrades. It is expected that changes to the CSL will be made to support each TI upgrade. For example, the level of I&T effort required to support a TI upgrade is half that for an ACB. Some portion of that effort is related to software changes required to support the new hardware. Second, incorporating software fixes and ACS element upgrades into each ACB and TI doubles the number of opportunities to incorporate software fixes. In combination with the CSL, this will rapidly improve the stability of the deployed combat system. Finally, offsetting ACB and TI development level-loads the Aegis technical infrastructure.

Inherent in the implementation of the IWS model is a development effort that consistently provides mature technology for integration in subsequent ACBs and TIs. This necessitates an understanding of the I&T impact of ACB and TI decisions on the Aegis technical infrastructure. Figures 8.2 and 8.3 show the effects of our proposed ACB/TI implementation approach on the Aegis facility and manpower infrastructure. In developing these estimates, we assumed that a TI upgrade is equivalent to 40 percent of the ACB-08 effort and an ACB is equivalent to 80 percent. Offsetting the ACB and TI upgrades results in a relatively level loading of facilities and personnel at approximately 40 percent of the total available. This allows PEO Integrated Warfare Systems to consistently plan and conduct its development efforts. Coincident ACB and TI upgrades aggregate I&T efforts in the two

Figure 8.2
Projected Aegis Facility Usage Assuming Proposed ACB/TI Implementation

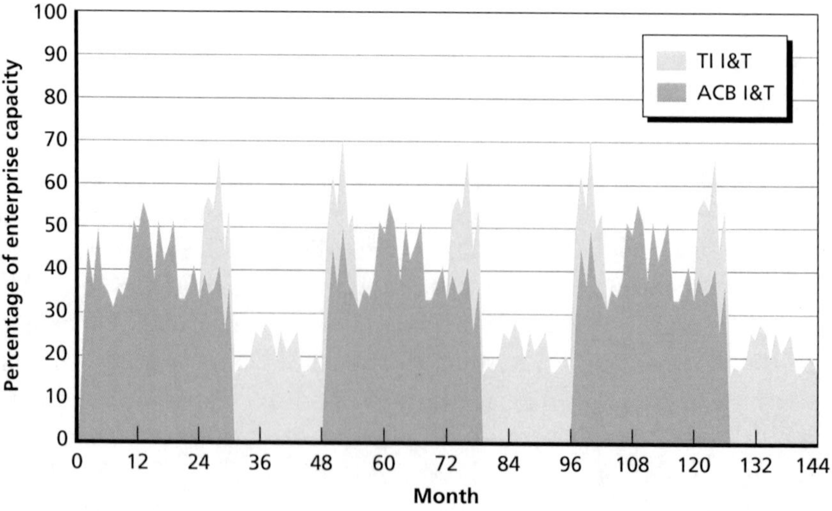

RAND *RR161-8.2*

Figure 8.3
Projected Aegis Technical Manpower Requirements Assuming Proposed ACB/TI Implementation

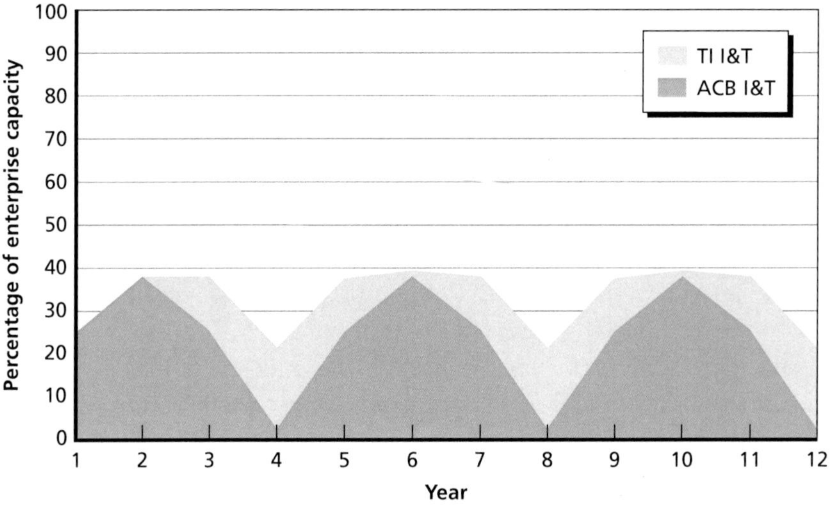

RAND *RR161-8.3*

years prior to deployment. This would require significant movement of people and facilities between integration and development.

Benchmarking the RAND Proposal

Several published proposed ACB/TI implementations are viable candidates for PEO Integrated Warfare Systems to consider. Under its IWS business model, the PEO proposes four-year coincident ACB and TI drumbeats, with individual ships receiving every ACB upgrade and every other TI upgrade. The U.S. Navy submarine community's ARCI program operates with an offset two-year APB and TI drumbeat, with individual ships receiving every other upgrade. When considering the utility of the ARCI program parameters to the Aegis program, it will be important to consider the caveats discussed in detail in Chapter Six.

In our analysis, we considered both the costs and the benefits of three potential program implementation approaches.[1] Table 8.1 details the impact of the three approaches on the fleet, as well as some of the benefits. To determine the impact on the fleet, we first considered the number of upgrades that the fleet must support and the number of deployed ACB and TI combinations in the fleet. In terms of impact on the fleet, all three plans entail the same number of ships (approximately 17) to receive an ACB upgrade annually. The ACBs provide the major capability upgrades and have the most signifiant impact on the ships' operators. However, from the perspective of availability, the ACB upgrades only minimally affect individual ships. These upgrades concern software only and can be accomplished quickly. The IWS model requires substantially fewer TI upgrades than ACB upgrades. That is as expected; individual ships receive a TI upgrade every four years under both the ARCI and RAND implementation models and every eight years under the IWS model.[2] The TI upgrades do not provide capability improvements but have a more significant impact on fleet availability. Individual upgrades replace all commercial computer hardware on a ship and require longer periods of availability.[3] The last fleet impact metric is the number of deployed ACB-TI combinations. The IWS and ARCI models result in approximately five deployed ACB-TI combinations. This represents a slight increase over the number of baselines deployed under the legacy business model. Our implementation approach decreases the number of combinations to fewer than three, on average. It has been suggested that Aegis ships with different baselines can have difficulty operating together.

The variations in upgrade frequency among the three models affect deployed Aegis capabilities. As discussed in Chapter Five, capability is measured by the age of the deployed technology for both software and hardware. The average software age is about four years for the

---

1 The results referenced in this chapter and in Table 8.1 assume steady-state statistics. This implies that the entire Aegis fleet has transitioned from legacy to OA combat systems.

2 The reduction under the IWS plan is more than a factor of two due to the legal limitations on upgrading warships in the final five years of life.

3 All upgrades can be completed within a four-week availability.

Table 8.1
Fleet Attributes of Various ACB/TI Implementation Schedules

Model	ACB Rate	ACB Lag	TI Rate	TI Lag	Upgrades per Year		ACB-TI Combinations in Fleet	Software Age		TI Age	
					ACB	TI		Max.	Avg.	Max.	Avg.
ARCI	2	2	2	2	17.1	17.1	4.6	5.3	3.8	5.3	3.9
IWS	2	2	4	2	17.1	7.7	5.5	5.3	3.9	10.1	6.8
RAND	4	1	4	1	17.1	17.1	2.8	6.1	4.9	6.5	4.9

IWS and ARCI models and lengthens to five years under our proposed model. When considering these variations, we note that the legacy business model results in an average software age of 14 years. Further, as described in previous chapters, a two-year ACB development time may not be feasible, given the complexity of the ACS and its difficult testing environment. Our model results in slightly higher software ages but doubles the available I&T time.

The average hardware age has more variability than the software age in the three models. The IWS business model and the ARCI model have average hardware ages of seven years and four years, respectively. Our model splits the difference with a hardware age of five years. We note that the IWS model significantly reduces the number of TI upgrades, so it is expected that the average hardware age will increase. On the other hand, the ARCI model minimizes technical risk by maintaining pace with the commercial computer industry's two-year drumbeat. The IWS model, in which individual ships retain hardware for eight years, has the highest level of implicit technical risk.

Implications for Development

As the Navy considers alternative upgrade intervals, it must balance upgrade size and frequency. Smaller, more frequent upgrades will distribute capability improvements across the fleet more quickly but will cause the fixed costs of I&T to consume more of the fixed IWS budgets. Larger, less frequent updates will distribute capability more slowly, but the fixed costs of I&T will consume less of the fixed budgets.

The choice of TI intervals requires the additional consideration of coordinating computing and networking hardware upgrades with the industry that produces the COTS equipment. The ARCI program disseminates upgraded hardware roughly every two years, which allows it to minimize procurement and in-service costs while leveraging recent, if not state-of-the-art, COTS equipment. Unfortunately, integrating new hardware in two years would be especially challenging for Aegis. However, the Navy may be able to mitigate the in-service cost of slower drumbeats by warehousing retired computing hardware and using the parts as spares. To date, it has not needed the computing capacity provided by faster upgrades.

Sources of Risk in the IWS Business Model

There are several sources of risk in the IWS business model. The most fundamental source risk, which should not be underestimated, is that the model is unlike the legacy approach to developing and fielding Aegis capability upgrades. The experiences of ARCI and SSDS provide some lessons learned, but Aegis is a fundamentally different program, and the Navy should expect to have an experience that is unique in complexity. Thus, the Navy should be prepared to harvest its own lessons as it proceeds to field periodic upgrades to modernized ships and to develop the system using a CSL.

There are other significant sources of risk as well, including the fact that multiple government stakeholders have a vested interest in the legacy business model; the complexity of managing a CSL; the possibility that the Navy and the MDA's BMD program will compete for limited talent, facility time, and access to the CSL; and the diversion of resources to the CAL from direct capability improvements. The Navy can mitigate some of these risks by making capital investments in the CSL and componentizing software, by delaying investments in product-line development until transition to the CSL is successful, by streamlining government involvement in I&T, by enforcing requirements discipline, by staggering TIs and ACBs, and by harvesting lessons learned from ARCI and SSDS.

Bibliography

Edwards, RADM M. J., "Requirement for Open Architecture (OA) Implementation," Memorandum, 9010 Ser N6N7/5U916276, December 23, 2005. As of October 10, 2011: https://acc.dau.mil/adl/en-US/31396/file/5659/OA%20Letter.pdf

Fein, Geoff, "Raytheon's SSDS Set for USS *Nimitz* Installation in '08," *Defense Daily*, December 4, 2007.

GlobalSecurity.org, "Aegis Ballistic Missile Defense (Aegis BMD)," web page, last updated July 7, 2011. As of October 10, 2011: http://www.globalsecurity.org/military/systems/ship/systems/aegis-bmd.htm

Hicks, RADM Alan B., "Aegis Ballistic Missile Defense: Aegis Ballistic Missile Defense Overview for the George C. Marshall Institute," briefing, Missile Defense Agency, August 3, 2009.

Jacobus, Peter W., Puck-Fai Yan, and John Barrett, "Information Management: The Advanced Processor Build (Tactical)," *Johns Hopkins APL Technical Digest*, Vol. 23, No. 4, October–December 2002, pp. 366–372.

Kaufman, Stephen, "Romania Agrees to Host Ballistic Missile Interceptor," *America.gov Archive*, February 4, 2010. As of October 10, 2011: http://www.america.gov/st/eur-english/2010/February/20100204155405esnamfuak0.8593866.html

Kerr, Gibson, "COTS: We Can't Afford to Do It Any Other Way," *Proceedings*, Vol. 132, No. 10, October 2006, pp. 68–71.

Miller, Greg, "Capabilities of COTS vs. MILSPEC Computing," briefing to the Program Executive Office for Integrated Warfare Systems, August 2010.

Missile Defense Agency, *DOT&E FY 2002 Annual Report: Ballistic Missile Defense System (BMDS); Space-Based Infrared System (SBIRS)*, February 21, 2003.

Naval Sea Systems Command, "Aegis Technical Representative: Mission," web page, undated. As of November 4, 2011: http://www.navsea.navy.mil/AegisTechRep.aspx

Naval Surface Warfare Center, Port Hueneme Division, "USS *Bunker Hill* (CG-52) Combat System Ship Qualification Trial Test Plan, Volume 1: Introduction," April 1, 2009.

Office of the Secretary of the Navy for Research, Development and Acquisition, *Guidebook for Acquisition of Naval Software Intensive Systems*, version 1.0, September 2008.

O'Rourke, Ronald, *U.S. Navy Force Structure and Shipbuilding Plans: Background and Issues for Congress*, Washington, D.C.: Congressional Research Service, RL32665, January 6, 2012.

"PEO IWS and NSWCNUWC FY10 Task Review: NSWC Crane Division," briefing, September 9–10, 2009.

Pfleeger, Shari Lawrence, and Joanne M. Atlee, *Software Engineering: Theory and Practice*, 3rd edition, Upper Saddle River, N.J.: Prentice Hall, 2005.

Program Executive Office for Integrated Warfare Systems, U.S. Navy, *Surface U.S. Navy Combat Systems Development Strategy Acquisition Management Plan (AMP)*, version 1.0, December 18, 2009.

———, *Surface U.S. Navy Combat Systems Development Strategy Acquisition Management Plan (AMP)*, 2013.

U.S. Department of Defense, "Ballistic Missile Review," briefing, February 2, 2010.

U.S. Department of the Navy, *Fiscal Year (FY) 2010 Budget Estimates: Justification of Estimates, Other Procurement, Navy Budget Activity 4*, May 2009.

———, *Report to Congress on Annual Long-Range Plan for Construction of Naval Vessels for FY2011*, Washington, D.C., February 2010a.

———, "Cruisers—CG," fact sheet, September 17, 2010b. As of October 10, 2011.
http://www.navy.mil/navydata/fact_display.asp?cid=4200&tid=800&ct=4

———, *Report to Congress on Annual Long-Range Plan for Construction of Naval Vessels for FY2013*, Washington, D.C., April 2012.

U.S. Navy Office of Information, "Open Architecture (OA)," *Rhumb Lines*, December 12, 2006.

Zarnich, Robert, "ASW Enterprise Open Architecture," briefing, February 21, 2006. As of October 10, 2011:
https://acc.dau.mil/adl/en-US/44946/file/13047/Presentation%207_ASW%20OA%20Overview.pdf